Cladistics

Cladistics and phylogenetic reconstruction are subjects which biology students find quite difficult to grasp when taught from conventional textbooks. This CD provides students with a complete self-study introductory course in phylogenetic reconstruction using cladistic analysis. The CD is fully interactive and includes animated sequences, questions at the end of each section, and practical exercises. It is the first exclusively pedagogical CD-ROM devoted to this topic. By the end of the course students should have a basic understanding of cladistics and be able to reconstruct phylogenetic relationships from morphological and molecular data.

The CD-ROM is accompanied by a short textbook. The book is meant to be used in conjunction with the CD-ROM but can act as a stand-alone aid to learning when the reader is away from the computer.

Cladistics
A practical primer on CD-ROM

by
Peter Skelton and Andrew Smith

Accompanying booklet by Neale Monks

CAMBRIDGE
UNIVERSITY PRESS

PUBLISHED BY THE PRESS SYNDICATE OF THE UNIVERSITY OF CAMBRIDGE
The Pitt Building, Trumpington Street, Cambridge, United Kingdom

CAMBRIDGE UNIVERSITY PRESS
The Edinburgh Building, Cambridge CB2 2RU, UK
40 West 20th Street, New York, NY 10011–4211, USA
477 Williamstown Road, Port Melbourne, VIC 3207, Australia
Ruiz de Alarcón 13, 28014 Madrid, Spain
Dock House, The Waterfront, Cape Town 8001, South Africa

http://www.cambridge.org

First published 2002

Printed in the United Kingdom at the University Press, Cambridge

Typeface Bembo 11/14 pt.   *System* QuarkXpress   [TB]

A catalogue record for this book is available from the British Library

ISBN 0 521 52341 9

Contents

Contents

Preface

The last few decades have seen a revolution in the reconstruction of evolutionary relationships. Eclectic models based on *ad hoc* arguments, and hence liable to subjectivity, have been replaced by a consistent methodology – *cladistics* – that is open to objective evaluation. Two technological innovations have helped to foment this revolution. First, computing, especially the rise of personal computers, has made easy the previously unthinkable task of sifting through myriads of alternative trees of shared ancestry, as required by the method. Second, molecular sequencing of genes and their products has provided a rich new source of evidence that is also amenable to cladistic analysis, in addition to conventional morphological data. Where these two discrete sources of data yield the same phylogenetic conclusion, a high degree of confidence can be placed on it. Thus what before was virtually a priestly art practised by taxonomic specialists has become a robust science.

Consequently, discussion of cladistics forms a necessary component of any modern course on evolution. However, many students have difficulties getting to grips with the strict logic of the method, not to mention its somewhat daunting technical vocabulary. Accordingly, the *Evolution Course Team* at the Open University decided in 1998 that treatment of the topic could especially benefit from the lively and colourful medium of an interactive CD-ROM, and thus the present project was born. A proposal for the structure and content of the CD-ROM was commissioned from Andrew Smith at the Natural History Museum, and this was realised in detail by Peter Skelton at the Open University, working with a technical design team from BBC Factual & Learning (MK). Co-publication was agreed with Cambridge University Press, who commissioned the present booklet from Neale Monks – an inspired choice – to accompany the CD-ROM. We, the authors of the latter, are indeed most grateful to Neale for the elegant but down-to-earth guidance notes and backup information that he has provided here for users of the CD-ROM.

Peter Skelton
Open University, Milton Keynes
Andrew Smith
Natural History Museum, London

About the CD-ROM and the booklet

The CD-ROM can be studied on its own. All the information you need to successfully complete this Course is on the CD-ROM, with exercises and a glossary of terms. However, if you feel the need for a 'guiding hand', then the booklet will take you through the CD-ROM section by section. There is nothing important in the booklet that isn't on the CD-ROM, or vice versa, but there are more detailed explanations of some aspects, as well as extra examples of some of the paradigms discussed, which might be helpful.

The multimedia presentation on the CD-ROM is divided up into a series of screens, which the sections of the booklet parallel. The first four sections contain the material you will need to learn to fully understand the Course, and we expect each of these sections to take 1 to 1.5 hours to complete. The final section is a series of practical exercises that test your understanding of the issues and techniques introduced here, and you should expect to spend 5 hours on this (though obviously you don't need to do them all in one go!).

Finally, a personal word from the author of the booklet. When I was asked by Peter Skelton and Andy Smith to write this booklet it was not because I was an expert in the methodology of cladistic phylogenetics, but because I was just a 'regular guy' who uses the methods described here on a day-to-day basis to solve evolutionary and taxonomic problems. Cladistics appears at first glance to be an arcane process with its own rich vocabulary of technical terms and buzzwords, but in fact the basic methods are very straightforward. We hope this will become apparent by the time you finish this Course.

Instructions for installing and running the CD-ROM

Macintosh

1. Put the CD-ROM into the computer.
2. Double-click on the **Start** icon to begin.
3. Click on the **Using this CD** button for a brief tutorial (note that, unlike most Macintosh applications, a single, not double, click is required).

Windows

1. We suggest that you will need at least 32 megabytes of RAM on your computer to run this CD-ROM. Please note that it contains spoken commentary, so ensure that the sound on your computer is switched on.
2. For optimal viewing, check that your machine is set up to display at least 640 by 480 pixels in at least 16-bit colour. You can do this by opening the **Control Panel** folder (which you can find under **Start – Settings**), double-clicking on **Display**, and then clicking on the **Settings** tab shown there. Then, if necessary, adjust the display area and colour as advised above. You may wish to experiment with different settings, as these affect the size of the CD images that you will see on the screen.
3. Put the CD-ROM into the computer and double-click on the icon for **My Computer**, then on the CD-ROM icon (which may be labelled **(E)**). This will open the CD-ROM program folder.
4. Launch the program by double-clicking on the file called **Cladistics.exe** (ignore the other folders shown there).

5. When the menu screen appears, you may wish first to have a look at **Using this CD** (icon at the bottom of the screen) before beginning your study of the main contents. Incidentally, please note that some of the exercises advise you to print off a given screen image; you can do this by clicking on the **Print** button under **functions**. You can use either a colour or a black and white printer for this.

1–First principles

1.1 – Reconstructing evolutionary history from observed differences

Screen 1/16

If you could look back at the evolutionary history of horses, zebras and cows through time, you would be able to see a series of branching events – moments when one species splits into two. Some of these branches might not lead to horses, zebras or cows, perhaps going extinct at some point, or giving rise to some other animal we haven't considered, such as donkeys or buffalo. But other branches would, and so it would be possible to trace the line of descent from the common ancestor of all three species to each species in turn. Looking at each branching event, we would see that each time one species gave rise to another, the characteristics of the descendent species were different in some significant way to its ancestor. The awkward thing for biologists is that we don't have nice detailed family trees – phylogenies – like this. So how can we reconstruct phylogenies and understand the relationships between taxa? The answer is by looking at the distribution of characters among the taxa being studied.

In Text-fig. 1, the three animals shown on the screen – a horse, a zebra and a cow – are shown together with a few characteristics that they have in common and others that they don't. Those characters which they all have in common might be good evidence from which to deduce that all three animals are closely related. For example, all three have hairy bodies, and all produce milk for their offspring. In fact, these similarities are typical for all mammals, the large group to which not only these three species

1

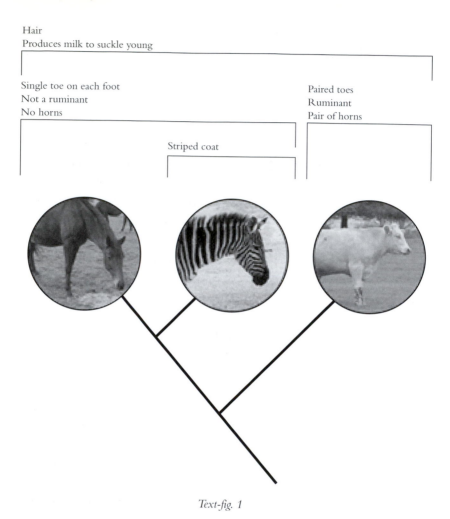

Hair
Produces milk to suckle young

Single toe on each foot
Not a ruminant
No horns

Paired toes
Ruminant
Pair of horns

Striped coat

Text-fig. 1

but also many other animals, including ourselves, belong. On the other hand, there are differences between horses, zebras and cows. Only cows have horns, and only cows 'chew the cud', i.e., are ruminants. Horses and zebras differ again from cows in having only a single toe, or hoof, on each foot, while cows have cloven hooves with two toes. Zebras differ from horses in being striped while horses are plain.

So, we can see that for these animals there are some things they all have in common and some they do not. At its most simple, reconstructing phylogeny is all about how you relate the presence or absence of characters among a group of taxa to their evolutionary history. In other words, characters can be thought of as the 'markers' that denote relationships. The fact that I have more of these markers in common with my father than with my grandfather reflects the fact that I am more closely related to my father than to my grandfather. The further back in my family tree you go, the fewer of these markers, or characters, I will have in common with that relative. Ultimately, of course, all humans are related and descended from the first early humans of Africa, and so we are all part of one big human family tree. Even so, you and I may look very different – our respective branches of the human family tree diverged generations ago. In exactly the same way zebras and horses share more characters than either do with the cow simply because zebras and horses diverged relatively recently (a few million years ago), while their common lineage split off from the one leading to the cow much further back (tens of millions of years ago). We can say that while zebras and horses shared a common ancestor recently, the common ancestor of zebras, horses and cows is much more ancient. On the family tree (or phylogeny) of the horse, the zebra is a close relative while the cow is a much more distant cousin.

Now that we can relate characters to phylogeny let us reflect a moment on what sorts of things can be used as characters. There are two main sources of characters: those determined from the bodily attributes of an organism, and those taken from the genes. The bodily attributes of an organism include its morphology, physiology and behaviour, and characters can be taken from any of these sources. An example of a morphological character is one that we have already mentioned, the presence of hair. This is a character shared by horses, zebras and cows. It is worth noting that palaeontologists in particular have to use morphological characters almost exclusively, since the fossil record reveals very little else. Biologists working on living organisms, on the other hand, can use a much broader range of character sources. The physiology of an organism can be very informative; for example, photosynthesis is a physiological process characteristic of plants but not seen in animals. Yet another source of characters is behaviour, although this can be problematic inasmuch as behaviours are much more flexible than morphology or physiology; animals can often adapt existing behaviours or learn new ones in response to changes in the environment. But apes bare their teeth – or smile – in just the same way as a human might do when nervously dealing with others, to indicate an absence of hostility. Dogs and cats bare their teeth – or snarl – as a threat. So, the fact that humans and apes share this behaviour, while dogs and cats do not, could be used as evidence for their close relationship.

Genetic characters are based on the sequences of the nucleotide bases adenine, cytosine, guanine and thymine (A, C, G and T) in the DNA molecule. Certain parts of the DNA code for specific proteins, and these parts are called *genes*. Each such gene will be made up of a certain number of bases and it is possible to identify which bases occur and write them out, a procedure called *sequencing*. Comparisons of a certain section of one gene in one organism with the same section of the same gene in another organism can reveal differences in the series of bases, and these differences can be used as characters for determining phylogeny. Genes code for amino acids, which are put together to make proteins, the essential 'building blocks' of life. They are fundamental to the way an organism develops and functions. Changes in the series of bases change the way the gene expresses itself, and ultimately the morphology, physiology and behaviour of the organism. Furthermore, genes are the stuff of heredity; an organism does not inherit the morphology, physiology and behaviour of its parents fully formed, but as the genetic code in its cells. As it grows, these genes express themselves and the cells take on the form and processes that these genes dictate. To take an obvious example: an acorn does not have any kind of leaves; only as it grows, and its shape develops, does it acquire the characteristic oak tree-shaped leaves. Inside the acorn were not miniature leaves waiting to emerge, but the genes to tell the dividing cells how to become leaves.

Screens 2/16 to 11/16

These screens relate changes in the series of nucleotide bases to the phylogeny of a group of organisms. At each successive time interval the series of bases possessed by the descendent species differ from those of its ancestor. These changes are called *mutations*.

Mutations of the genes are the essential mechanism by which lines of descent change, and are fundamentally different to accidental changes to the morphology, physiology and behaviour of an organism. Take an example of a boy who breaks his leg. The break results in a change to his morphology. Although after a while the leg will heal and will work normally, the leg bone itself will be slightly different to how it was before, having a new mass of bone around where the fracture was, rather like a scar. But when the boy grows up and has children, do they also have leg bones like his? They do not; their bones are perfectly normal. This is because while his leg bone was damaged, the genes that show how to make a leg bone are fine. So his sons and daughters inherit the normal genes for making leg bones.

Mutations to genes, on the other hand, are inherited. If an organism acquires some sort of genetic mutation, then its offspring can inherit that mutation. Mutations can happen for all sorts of reasons, and are a natural part of life. Indeed, they are essential to life being able to change and adapt through natural selection. Sometimes mutations are *lethal*: they prevent an essential gene from working properly, and prevent the organism from reaching maturity. If an organism cannot live long enough to have offspring, then it has no chance of passing on the mutation to the next generation. In this way, natural selection very swiftly wipes out debilitating mutations. Other mutations do not harm the organism, but neither do they confer any benefits; these are called *neutral* mutations, and over time they can become fixed in an evolving lineage simply by a process of random drift. Finally there are useful, or *adaptive*, mutations that give an organism some competitive edge over its peers. An organism that can do something beneficial to its survival is an organism that has a better chance of reaching maturity and of having more offspring. So this time natural selection works in the other way, favouring the inheritance of this new, mutated gene by the next generation over the original version of the same gene. Over time, the entire population of an organism could end up sharing this mutation. The mutation could be one that causes some new morphological, physiological or behavioural trait, and so all the descendants which share this mutated change will also share this trait.

The hypothetical phylogeny considered in these screens begins at time T_0 and ends at time T_3. At each time interval the series of bases possessed by each organism is listed on the right-hand side. There are twenty bases in total. Beginning at time T_0 (Screen 2/16), there is only a single species, represented by the green circle on the line. Its genetic sequence is shown on the right. Advancing to the next screen (Screen 3/16) causes time to pass, and we can watch a series of mutations occur. The first mutation is inherited by all the members of the species, a change from T to C, as shown in Text-fig. 2.

After that mutation comes a branching event. One part of the population inherits a mutation from C to T, while another inherits a mutation from A to G. This is shown in Text-fig. 3. Note that these mutations occur exclusive to one another, i.e., each mutation characterises each branch, and the population making up each branch is now genetically

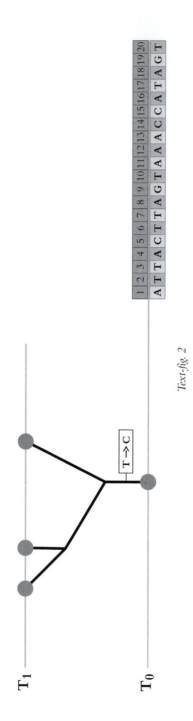

Text-fig. 2

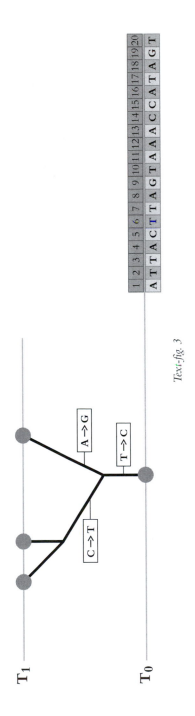

Text-fig. 3

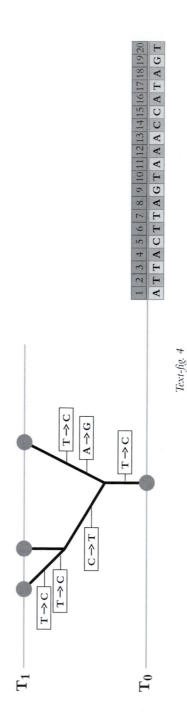

Text-fig. 4

distinct not only from the ancestral population but also from the population making up the other branch.

As time goes on, further mutations take place, as shown in Text-fig. 4. On the right-hand branch, a second mutation occurs. This is inherited by all members of the population of that branch, so that the circle at the top of the right-hand branch, i.e., at time T_1, is now a single species sharing both the A to G and T to C mutations relative to the ancestral species. The left-hand branch is different, because it divides a second time to form two distinct species, i.e., the two circles at time T_1. One of these species is made up of a population that has inherited not only the original C to T mutation but also two further mutations, in this case mutations from T to C at two different points in the gene.

Screens 4/16 to 7/16 summarise this. Run through them carefully and observe how the mutations plotted on the branches correspond to the changes in the data matrix on the right-hand side. Look at Text-fig. 5 and note, for example, that the base position 2, which was originally T, becomes C and that all three descendent species inherit that condition. By contrast, base position 5, originally a C, is changed only in the branch leading to the two taxa on the left, where it becomes a T.

As the slideshow progresses, more time passes, from T_1 through T_2 and finally to T_3. Watch as further mutations and branching events take place, resulting in six species at time T_3. Screen 9/16 allows you to look at each base position in turn. If you hold the cursor over one base position in the data matrix, all the corresponding bases for all the species are highlighted in red. This makes it easy to see the changes that affect that particular point in the gene. If you look at base position 1, for example, you will see that all species have an A here. No changes have occurred. Base position 2, on the other hand, changes once from a T to a C between time T_0 and T_1, but never again. All descendent species share this condition. On the tree which mutation was this?

To make things a little clearer, in Text-fig.6 the ancestor has been labelled **X** and the descendent species labelled with letters **a** through **n**. If you look at this figure, you can see that base position 2 corresponds to the mutation at the very bottom of the tree, between X and its descendants **a**, **b** and **c**. Since it occurs before the diverging of **a** and **b** on the one side and **c** on the other, both branches inherit this mutation.

Now try base position 12. This mutation happens twice, independently, once in the branch leading to **j** and **k** during the interval T_2 to T_3,

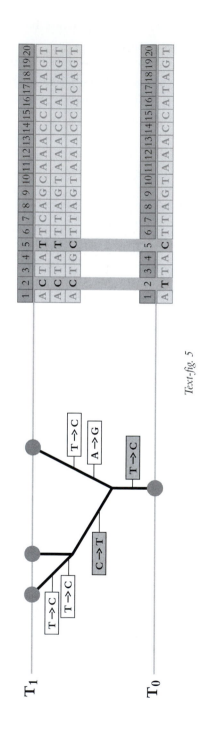

Text-fig. 5

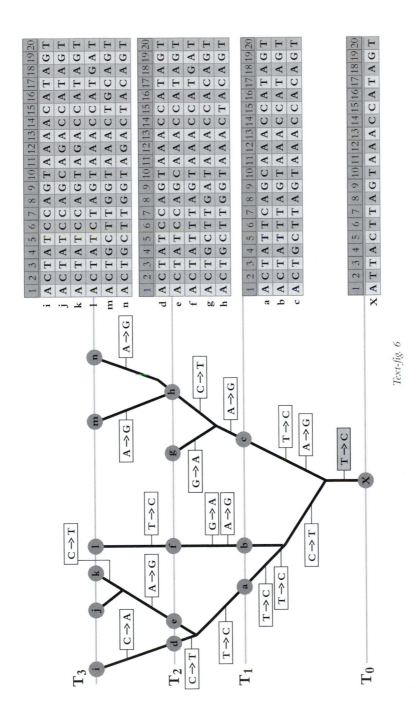

Text-fig. 6

11

and again in the branch leading to **n**, during the same time interval. This is shown in Text-fig. 7.

It is the identification of such changes that serves to identify clades, as is shown on Screen 10/16. Each clade shares one or more base changes, and such changes are said to be shared, derived features. Cladistics is the process of identifying such shared changes and using them to understand the relationships between taxa.

Screens 12/16 to 14/16

Now we are going to see how knowing the characters – the base nucleotides for the twenty base positions – of three descendent species, called **x**, **y** and **z**, and the same characters for the ancestor, **w**, can help us to establish the evolutionary history of the whole group; in short, we are using what we know about the species to deduce their phylogeny.

Screen 12/16 reduces the tree that we have been looking at to just three descendent species and the ancestral taxon. At first glance it might seem that we have lost all the rest of the information. So how can we answer the central question of phylogeny, i.e., which two species are more closely related to each other than to the third? In this example there are three possible answers: **x** and **y** are more closely related to each other than either is to **z**; **y** and **z** are more closely related to each other than either is to **x**; or finally **x** and **z** are more closely related to each other than either is to **y**.

The solution is to establish which two descendent species share the greatest number of shared, derived characters. First, discard the base positions which are the same for the ancestor and all the descendants, since such characters can give no useful information about the phylogeny of the group. You can think of them as being too undiscriminating at this level of analysis, in the same way as presence or absence of hair wouldn't be a useful character in working out the phylogeny of a horse, zebra and cow. Such characters are base positions 1, 3, 9, 11, 13, 14 and 20. Equally unhelpful are characters in which all three descendent species share a different state to the ancestor, for much the same reasons. Base position 2 is such a character. Finally, we can ignore characters which are different in only one of the descendent species, the other two retaining the same character state as the ancestor. This is what we can see for base positions 4, 7, 8, 10, 12, 15, 16, 17, 18 and 19. Such characters can't be used to reveal a close relationship between two species as only one species has the change

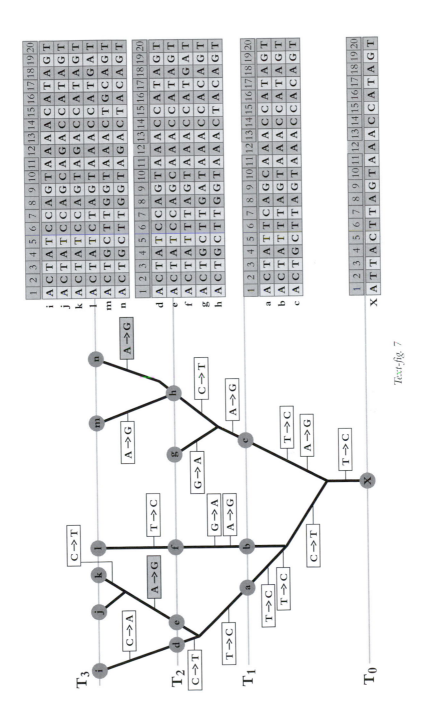

Text-fig. 7

13

or mutation, i.e., it is not shared. In fact the only characters which occur in the changed, or derived, state relative to the ancestor for two of the three descendent species are base positions 5 and 6 (Screen 13/16). Species **x** and **y** share a T at base position 5 and a C at base position 6, while **z** retains the same bases as the ancestor at these positions (C and T, respectively). So **x** and **y** share two derived characters relative to **z**.

Which tree best fits this hypothesis? Biologists assume that by describing an evolutionary hypothesis that explains all the observed character states with the smallest number of steps, they are minimising the amount of conjecture, i.e., that the simplest tree is the least speculative. Essentially this is saying that a phylogeny which implies the fewest number of changes is the most likely, and intuitively this makes sense. It is much easier to believe that ducks, storks and eagles all have wings because their shared common ancestor had wings than that their common ancestor had no wings and that each separate lineage independently evolved them. This is the *principle of parsimony*.

The tree that fits this hypothesis is the first one. The change for base position 5 from C to T and for base position 6 from T to C need happen only once at the base of the branch containing **x** and **y**. For the other trees we would have to infer more changes, as shown here in Text-fig. 8. Clearly while the first tree needs only two changes, the other two need four: the first tree is said to be the *most parsimonious*.

By comparing the deduced tree with the original phylogeny we can see that they are compatible. Thus using a parsimony analysis we have been able to recover some valuable information about the phylogeny of the group simply by looking at the distribution of characters of a few of the species.

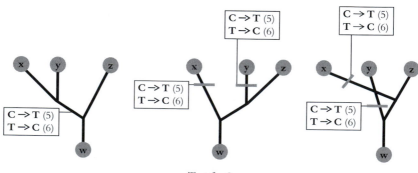

Text-fig. 8

Screens 15/16 and 16/16: The Three Species Game

This exercise tests everything we have looked at in the previous screens, and gives us the opportunity to pull together what we have learned to perform a cladistic analysis much like those done by biologists. Again, there are twenty characters, base positions on a gene, numbered 1 to 20. Compare each of these characters for each descendent species (**x**, **y** and **z**) with the ancestor, **w**.

Once again, the first thing to do is to find out which characters are helpful and which ones are not. Remember, characters in which all the descendent species show the same state, whether primitive or derived, are uninformative. They don't tell us anything about the relationship between **x**, **y** and **z**. That means that characters 1, 2, 3, 9, 10, 11, 13, 14, 18, 19 and 20 are uninformative. For all practical purposes, we can ignore them. Next we have to weed out characters for which only a single species has a derived state relative to the ancestor. Characters 5, 6, 7 and 16 are of this type. Again, they can be ignored.

Informative characters are ones where two taxa share a derived state which the third does not. So, if **x** and **y** shared a character state that **z** did not, then that character might suggest that **x** and **y** are more closely related to each other than either is to **z**. If, on the other hand, **y** and **z** shared a character state that **x** did not, then **y** and **z** might be more closely related than either is to **x**. Finally, a character state shared by **x** and **z** but not **y** would indicate a closer relationship between **x** and **z** than either has with **y**. In fact the only such characters in the data matrix are 4, 8, 12, 15 and 17, and these are the correct ones to highlight on the matrix.

With regard to phylogeny, what can we deduce from these characters? If we put the taxa which share the same state for any given character in parentheses, and leave the taxon with a different character state outside, we get these results:

4	$x, (y + z)$
8	$x, (y + z)$
12	$(x + z), y$
15	$x, (y + z)$
17	$x, (y + z)$

Of these informative characters we can see that four suggest that **y** and **z** are more closely related to each other than either is to **x** (characters 4, 8, 15 and 17). Only a single informative character, character 12, suggests a different relationship, i.e., that **x** and **z** are more closely related to each other than either is to **y**. So, a 4 to 1 majority favours the relationship **y** and **z** are more closely related to each other than either is to **x**. The tree which matches this hypothesis is the second one, and so that is the correct one to select.

1.2 – Parsimony and tree reconstruction

Screens 1/21 to 3/21

This section uses some real organisms – a butterfly, a moth, a house fly and a crane fly ('daddy long legs') – to illustrate how phylogeny is deduced from morphological characters. On the first screen we see one possible network of relationships, with the butterfly and crane fly on the left, and the house fly and moth on the right. Each pair is forming a clade relative to the other, as shown in Text-fig. 9.

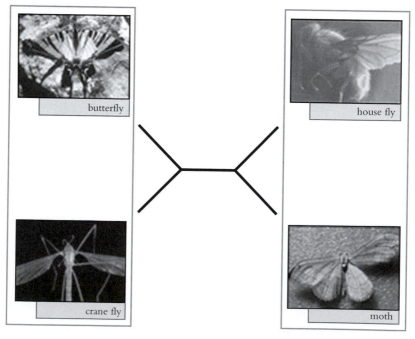

butterfly

house fly

crane fly

moth

Text-fig. 9

While this network says something about the relationships of the four species, it doesn't make any statements about ancestors or descendants. Such a cladogram is said to be *unrooted* – compared with the phylogenies mapped over time we looked at previously, this cladogram has no root, or putative common ancestor. (Go back and look at the phylogenies in Section 1.1. They describe the evolution of lineages over time, from a common ancestor to its various descendants. This unrooted cladogram is quite different.)

Another thing about unrooted cladograms is that the branching is always shown as a series of dichotomies, as shown on Screen 2/21. This is purely a convention: there is no fundamental reason why one species cannot simultaneously give rise to three or more species, though it is difficult to imagine such a thing happening in real life.

In fact the arrangement shown in Text-fig. 9 is by no means the only one possible. The animation on Screen 3/21 shows a second arrangement, with the butterfly and moth on the left-hand branch and the house fly and crane fly on the right-hand branch. How many others are there?

There are four species in total: the butterfly, the moth, the house fly and the crane fly; but that doesn't mean that there are four possible networks of relationships. Take the butterfly as an example. It could be:

1. More closely related to the crane fly than either the house fly or moth.
2. More closely related to the moth than either the crane fly or house fly.
3. More closely related to the house fly than either the crane fly or moth.

Likewise, if you frame these statements relative to any of the other species, you still end up with the same three possible arrangements. Since there are already two networks shown, the correct answer, then, to the question asked is one further tree, which the animation reveals after you have made your selection. If you look at the three arrangements listed above, you will notice that they match the three networks shown on the screen. The first arrangement matches the top cladogram, the second the next cladogram down, and the third the final cladogram.

Screens 4/21 to 10/21

Now we turn our attention to the principle of parsimony, and use it to deduce the phylogeny of the four insects we've been looking at – the butterfly, the moth, the house fly and the crane fly. To do this we need to look at some characters, to help us choose between the three possible networks.

The butterfly and the moth both have two pairs of wings, while the crane fly and house fly have only one pair of normal (flying) wings; instead of the second pair of wings these animals have highly modified 'halteres', balancing organs rather like gyroscopes. It is these that give flies their amazing agility. The wings of moths and butterflies are different from those of the house fly and crane fly. They are covered in microscopic scales, which if you have ever handled a butterfly or a moth you will recognise as the 'dust' that comes off on your fingers. On many butterflies and moths these scales are iridescent, and it is this that gives them their beautiful colours. The wings of the house fly and crane fly are simple, transparent sheets of tissue supported by veins. These characters can now be entered into a data, or character, matrix, as shown in Text-fig. 10.

On Screen 6/21 the three possible networks are shown again, next to the character matrix. The aim now is to use the attributes we can see on

	pairs of wings	scales on wings
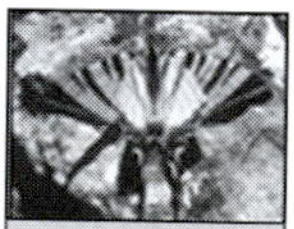 butterfly	2	Yes
 moth	2	Yes
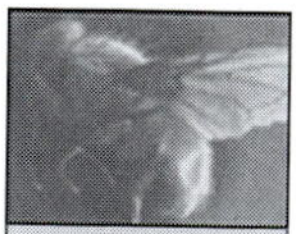 house fly	1 & 1 pair halteres	No
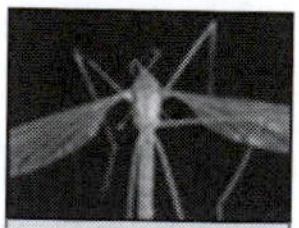 crane fly	1 & 1 pair halteres	No

Text-fig. 10

these four insects – the number of wings and the presence of scales on the wings – to determine which of the three networks is the most probable. A problem is that from these data alone it is impossible to know which version (or character state) of each attribute (or character) was present in the common ancestor. So we don't know whether the common ancestor of these four insects had two pairs of wings, or one pair plus the halteres. Equally, we don't know whether this ancestor had plain wings or wings with scales on. In cladistic terms, we cannot say which state is *primitive* and which is *derived* for either character. Even if we cannot know this, we can assess the number of character transitions each network demands.

Taking the top network as an example, consider the transition between two pairs of wings and one pair of wings plus a pair of halteres. On the one hand, it is possible that a pair of wings and a pair of halteres is the derived condition, and that the common ancestor of all four insects had two pairs of wings. In this case, the crane fly and the house fly must have independently acquired the derived condition of a pair of wings and a pair of halteres. On the other hand, it could be that the primitive condition was one pair of wings and one pair of halteres, in which case the butterfly and the moth must have independently lost the halteres and gained the second pair of wings. In both cases, two independent character transformations are needed to make the observed data fit the network.

Exactly the same thing happens if we look at the second character, the presence of scales on the wings. If scales were present on the wings of the ancestor, and so are the primitive character state, then they must have been independently lost by the crane fly and the house fly; but if the ancestor had plain wings, making scales the derived character state, then the moth and the butterfly must have independently gained them. Either way, two character transformations are needed.

However the observed character data are plotted on the first network, you end up with a total of four character transformations, two for each character. What about the next tree? Screen 8/21 asks what is the minimum number of character transformations. Look at the characters and try to plot them against the network. The left-hand clade, the butterfly and the moth, both lack halteres but have scales on their wings. So, that entire clade can be defined by those characters. Likewise, the right-hand clade, the house fly and the crane fly, both have halteres and one pair of wings, and do not have scales on their wings. The horizontal line joining the two clades can therefore be defined by only *two* characters – the acquisition of scales on the wings for the clade including the butterfly and moth, and the acquisition of halteres for the clade including the house fly and crane fly, or the reverse in each case.

Compare this result with what we get if we look at the third network. This network needs four character transformations. Just like the first network, two independent transformations between two pairs of wings and one pair plus halteres, and two independent transformations between scaly and plain wings, are needed. To summarise, then:

- Tree 1 = four character transformations
  1. No scales on wings to scales on wings = butterfly
  2. Two pairs of wings to one pair of wings + halteres = house fly
  3. No scales on wings to scales on wings = moth
  4. Two pairs of wings to one pair of wings + halteres = crane fly
- Tree 2 = two character transformations
  1. No scales on wings to scales on wings = (butterfly and moth)
  2. Two pairs of wings to one pair of wings + halteres + (house fly and crane fly)
- Tree 3 = four character transformations
  1. No scales on wings to scales on wings = butterfly
  2. Two pairs of wings to one pair of wings + halteres = crane fly
  3. No scales on wings to scales on wings = moth
  4. Two pairs of wings to one pair of wings + halteres 5 house fly

If we want to choose one of these three as being the most probable pattern of relationships using the principle of parsimony, we need to choose the one with the fewest transformations. Trees One and Three are four 'steps' long while Tree 2 is only two 'steps': therefore, the second tree is the most parsimonious.

Screens 11/21 to 14/21

Having chosen the most parsimonious network of relationships, the next step is to turn that network into a phylogeny – a hypothesis describing not just relationships but also evolutionary history. To do this, the network needs to be *rooted*, i.e., it needs to state the order in which the branches diverged. We could make the moth the most basal branch, which would re-draw the network as a *cladogram*, a rooted network of

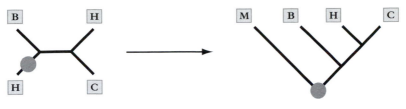

Text-fig. 11

relationships describing a phylogenetic hypothesis. One such cladogram compatible with the unrooted network of relationships that we have selected is shown in Text-fig. 11.

The divergence between the moth on the one hand and the remaining three species on the other is denoted by the circle. In fact, as Screen 12/21 shows, there are a further four ways of re-drawing the network of relationships as a cladogram, making a total of five. All five possible cladograms are topologically distinct, but compatible with the unrooted network. Likewise, if we consider the two other networks, the ones we had previously rejected as being unparsimonious, each of these has its own unique set of five distinct but compatible cladograms. So for these four species (or, for that matter, any four taxa of any kind) there are a total of fifteen possible cladograms. How do we choose between them?

For a start, if we are going to adopt the principle of parsimony as our criterion for choosing between them, we can dispense with the ten trees from the two unparsimonious networks at once. If the networks were unparsimonious, then so too will be the cladograms derived from them. But that still leaves five trees to choose between, as shown on Screen 14/21. To choose between them we need to know the direction evolution took between the two states of each character.

Screens 15/21 to 18/21

The direction of evolution is technically described as being from the *primitive* towards the *derived*. One must be aware that primitive does not necessarily imply the character state in question is crude or rudimentary, nor derived imply the character state is better or more sophisticated. The terms simply relate to which came first, i.e., the order in which they evolved.

Take for example the lungs in the land vertebrates (amphibians, reptiles, birds and mammals). From our perspective, lungs would seem to be an advanced character relative to the non-terrestrial vertebrates (the bony fish,

21

sharks and lampreys). Lungs are anatomically related to the swim bladders of bony fishes. Lungs and swim bladders develop in a similar way, as pouches from the gullet, and indeed some fishes, notably the lungfishes, bichirs and garpike, have lungs instead of a swim bladder. Just like land vertebrates, and unlike most other bony fish, these fishes can breathe air. The fossil record has shown that these fishes belong to the earliest groups of fishes, implying that lungs were a feature of the first bony fish, and have become modified over time to form a swim bladder. So rather than lungs being the advanced character state, they are in fact primitive relative to the development of the swim bladder in bony fishes, despite the fact that they are tremendously important and sophisticated structures essential to life on land.

A similar case applies when looking at structures which have become degenerate as a lineage evolves over time. Humans have a rudimentary tail, a few fused bones at the base of the spine. The tails of monkeys are clearly much more fully developed, particularly in those species in which the tail is prehensile. But that doesn't mean that the human tail is the primitive state for primate tails; quite the reverse is true in fact. Most living primates have well-developed tails, and the fossil record clearly shows that the degeneration of the tail is an advanced condition characteristic of the great ape clade to which humans belong.

The problem is that, *a priori*, it is very difficult to know which state is primitive and which is advanced. Observations of embryonic development and comparing various fossils are two widely used ways of determining which states are primitive and which are advanced (this is discussed in Section 3.1).

Returning to the phylogeny of the four insect species that we have been looking at, two statements about the characters used are made on Screen 15/21: first, that halteres were derived from the second pair of wings, implying that two pairs of wings is the primitive character state; and second, that the acquisition of scales on the wings is a derived character state relative to the primitive state of plain wings. Screen 16/21 maps these character transformations on to one of the cladograms, as shown in Text-fig. 12.

This cladogram states that the moth and the butterfly both independently acquired scaly wings (the dark grey bars), while the lineage including the house fly and the crane fly transformed the second pair of wings into halteres (the lighter grey bar). In other words, the house fly and the crane fly both inherited halteres from their common ancestor. Counting up these character state changes gives a total of three.

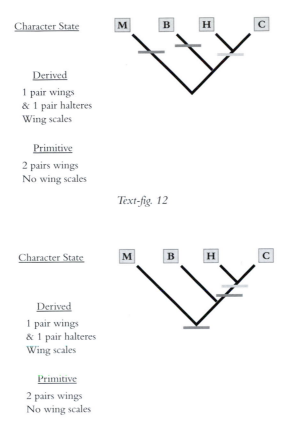

Character State

M B H C

Derived

1 pair wings
& 1 pair halteres
Wing scales

Primitive

2 pairs wings
No wing scales

Text-fig. 12

Character State

M B H C

Derived

1 pair wings
& 1 pair halteres
Wing scales

Primitive

2 pairs wings
No wing scales

Text-fig. 13

You could change things around and make the acquisition of scaly wings a characteristic of the common ancestor of all four insects: then the lineage leading to the house fly and the crane fly would need to lose the scaly wings, as well as transform the second pair of wings into halteres. Once again, there are three character state changes, as shown in Text-fig. 13. The loss of some feature by a taxon that was gained by the ancestors of that taxon is termed a *secondary loss*. This is a very different proposition from a taxon never having had that feature at any point in its evolution. For example, the lack of hind legs in whales is a secondary loss since their ancestors, land-dwelling mammals, had them; the absence of hind legs in sharks is not a secondary loss, as sharks never had them.

So, however you describe the tree in terms of character transformations, in total there need to be three transformations to make the cladogram fit the observed character data.

On Screen 18/21 we are asked to determine which of the five clado-grams contains the fewest character transformations (or steps). The top right cladogram and both the bottom cladograms all require three char-acter transformations, in the same sort of way as we have seen for the cladogram at the top left. Either one derived character state is acquired by the clade containing the top two species and each of the two bottom taxa independently acquire the other, or the entire clade acquires one of the derived character states, and the top two taxa then share a reversal of that character transition (a secondary loss), and also share a common acquisition of the second derived character state.

Only the middle cladogram is different, as shown in Text-fig. 14. It can be described by the acquisition of the derived state for one character by one clade, and the acquisition of the other derived state of the other character by the other clade, in this case by the acquisition of scaly wings by the clade including the moth and the butterfly, and the transformation of the sec-ond pair of wings into halteres by the clade including the house fly and the crane fly. Hence this is the most parsimonious solution.

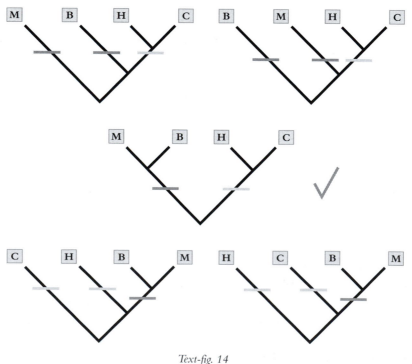

Text-fig. 14

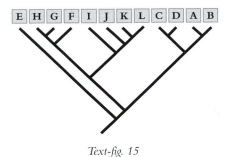

Text-fig. 15

Screens 20/21 and 21/21

Rooting Test 1 (Screen 20/21) presents us with a similar sort of problem. In the top left-hand side is an unrooted network of relationships. Admittedly, it is a good deal more complex than the four-insect example we have been looking at hitherto, but notice that all the branching is dichotomous. It is really nothing more than a number of four-taxon networks joined together. The easiest way to compare the cladograms with the unrooted network is to follow the network a junction at a time, and see if they match the topology of the cladograms. Let us compare the network with the cladogram at top right. If we look at the network first, we can see that **A** makes a dichotomous junction with **B** on the one side and everything else on the other. Moving up the network a step, we have **A** and **B** making a dichotomous junction with **C** and **D** on the one side and everything else on the other. You can make a series of these observations all the way along the network.

In fact you will observe that this particular cladogram, and the cladogram at bottom left, are compatible with the network. The cladogram at bottom right, however, is not; **I** and **K** are in the wrong places on the tree. To fit the network, the cladogram needs to be re-drawn as in Text-fig. 15. So, to answer this test correctly, choose the bottom right cladogram and highlight **I** and **K**.

Rooting Test 2 (Screen 21/21) is a very similar problem, and can be solved in the same way.

2–Characters and homology

2.1 – Homology and homoplasy

Screens 1/9 to 9/9

Though humans, dolphins and eagles might look very different, in many fundamental respects they are put together, so to speak, in the same way. These underlying structural similarities are called *homologies*, and are very important in determining relationships. Our arms, the fins of dolphins, and the wings of eagles all contain the same basic arrangement of bones, as can be seen on Screen 3/9, where the corresponding sets of bones have been colour coded. Note that although structurally similar in detail, superficially at least arms, fins and wings all look very different. They each serve different functions, and so have become morphologically adapted to different tasks.

The wings of insects and birds are not homologous, despite the fact that they serve the same purpose and function in a similar sort of way. Structurally they differ in virtually every detail: instead of bones, there are blood-filled veins for support, and there are no feathers over the skin, for example. Instead of an homology, we have an example of *homoplasy*, i.e, the superficial similarity of radically different structures. Birds and insects have each evolved wings to serve the same function – flight – but quite independently; birds' wings and insects' wings are functional analogues.

Striking examples of such analogues can be seen where different organisms have evolved to occupy a similar ecological niche, a process called *convergent evolution*. On the island of Madagascar there were, until domestic cats arrived with human settlers, no true cats of any sort, big or small. Occupying the big-cat niche on the island is the fossa, a mountain lion-sized civet. Elsewhere civets are small weasel-like animals, only distantly related to true cats. The physical and behavioural similarities between the fossa and a typical big cat are striking, and it takes a trained eye to distinguish them. Other good examples of convergent evolution can be seen between the swifts and the swallows, which are not closely related despite appearances, and between the succulents of Africa and the cacti of the New World.

Homology and homoplasy are relative terms, though. It all depends on the level of the phylogenetic analysis. In trying to understand the relationships of the phylum Animalia, i.e., all invertebrate and vertebrate multicellular animals, the fins of dolphins and a bony fish like a pike are sufficiently similar to be regarded as homologous structures. Containing supports made of bones and cartilage, they are both typical vertebrate appendages. As such, they are unlike the arms of the octopus or the legs of the spider. On the other hand, there are major differences in detail. Dolphins' fins have only a few bones embedded in the muscles and ligaments, and are capable of only limited movement. By contrast, the fins of bony fish contain dozens of fine bony rays connected by a flexible fin membrane. Such fins are very flexible and allow fish to use their fins not just for basic steering but also for sculling, tight turns and effective braking. So, as fins they are not very alike at all (Screen 5/9).

In fact the arrangement of bones seen in the fins of dolphins, the wings of birds and the arms of humans are characteristic of a group called the tetrapods, which includes the amphibians, reptiles, birds and mammals. By contrast, bony fish retain essentially the same sort of fin structure as did the common ancestor of the bony fish and the tetrapods. Bony fish can be said to exhibit the primitive condition for the vertebrate limb, while the tetrapods have the derived condition. Where one character state is believed to have been derived from another through the process of evolution, the primitive character state is called a *plesiomorphy* (from the Greek, meaning 'near form'), and the derived one an *apomorphy* (meaning 'beyond form').

Characteristics which define clades, as the tetrapod limb does the tetrapods, are termed *synapomorphies*. Such characters are extremely useful to taxonomists, acting like identity badges marking out members of clades even where other features might obscure their relationship. The

Greek philosopher Aristotle, for example, correctly deduced that dolphins were mammals as they produced milk and suckled their young at a time when most people regarded dolphins and whales as fish. In effect, he observed a synapomorphy of the mammals.

But how do we recognise valid synapomorphies? Screen 8/9 demonstrates one way, which is to compare one characteristic we suspect to be a synapomorphy with another character that might also be a synapomorphy. In this case the vertebrate limb with its internal skeleton and muscles is tested against the vertebrate eye. The taxa being looked at are a bony fish, a dolphin, a human, a bird and a house fly. Let us consider the limbs first. The fish, dolphin, bird and human all share limbs with an internal skeleton, connected by external muscles and ligaments, and covered by skin. The house fly has wings and limbs put together in a totally different way. Its wings are made from a sheet of tissue supported by veins, and its legs have an external skeleton with the muscles running along the inside. So, limbs at least do seem to support unity of the fish, dolphin, human and bird clade relative to the house fly *outgroup*.

Now look at the eyes. All four vertebrates have an eye consisting of a single chamber (the eyeball), the inside of which is lined with light-sensitive cells (the retina). A transparent lens at the opening focuses the light on the retina to form a sharp image. The house fly has an eye made in a different way. Instead of a single lens and retina, its eye contains thousands of small lenses each focusing on separate light-sensitive cells. Such an eye is called a compound eye, and it cannot form a particularly sharp image, though such eyes are very sensitive to movement. Once again it seems that the house fly can be excluded from the vertebrate group, this time on the basis of the structure of its eyes.

These two characters – limbs and eyes – produce the same pattern of relationships, i.e., that the fish, dolphin, human and bird are more closely related to each other than to the house fly. These characters are *congruent*. This is the ultimate practical test for any putative synapomorphy: the more characters support one another in defining a particular clade, the more probable it is that they are valid synapomorphies. In the example here, if we postulated that wings were a synapomorphy of a clade including the bird and the house fly, we would find that such a clade would not be supported by any other characters. So our hypothesis that wings were a synapomorphy of such a clade would fail. Instead, a character such as the possession of wings would be seen to be an example of homoplasy.

2.2 – Homology in molecular data

Screens 1/8 to 8/8

With morphological characters, it is relatively easy to make a good guess as to which ones are homologous and which ones are not. The fins of dolphins and the wings of birds occur in the same place relative to the rest of the skeleton, so their positional similarity is a clue to their common structure and development. Similarly, with homologous physiological or behavioural characters there is often the same sort of correspondence; a certain enzyme will be produced by the same organ, say. Unfortunately, molecular data do not behave in the same way.

Within the genome – the entire genetic 'blueprint' of the organism – genes do not necessarily exist as a single neat series of bases. Often they are divided up into sections in different parts of the genome, similarly to the way a single document on a computer might be broken up into several pieces on the hard disk. Functionally it is a single unit, but physically it is divided into several fragments all occupying different places. Complicating things still further, different species, even closely related ones, can have different numbers of chromosomes, so that the distribution of genes among the chromosomes can vary.

Instead of simply looking for the same genes, then, taxonomists look for similar sequences of nucleotide bases. Even if the gene itself is divided up, there should still be significant sequences of bases which can be compared with sequences of bases in other species. The more sequences match, the closer the relationship between the species is inferred to be. Screen 2/8 shows how this works. The two rows of nucleotide bases are taken from two species, Taxon 1 and Taxon 2. A particular twelve-base section from Taxon 1 is being compared with Taxon 2. Initially, only four out of the twelve bases match those of Taxon 1, a 33% match. If you scroll the series to left and right, you are in effect moving the twelve-base section from Taxon 1 along the entire length of the sequence of bases from Taxon 2, looking for a better match. Sometimes fewer bases match, and sometimes more. You should eventually find that a 100% match is possible – a perfect correspondence between the particular sequence from Taxon 1 and a certain section of Taxon 2.

Of course evolution affects genes, and so over time diverging lineages acquire differences in the sequence of bases in their genes. Screen 3/8 represents two taxa which have diverged somewhat, so we cannot expect

AGGATGAATCCC
G C A **AGGATAATCCCC** A T A

Text-fig. 16

AGGATGAATCCC
G C A **AGGATAATCCCC** A T A

Text-fig. 17

a 100% match. Though there should be a good match at some point with the twelve-base section from Taxon 1 and some section of the series of bases from Taxon 2, some of the bases in Taxon 2 might have been substituted with others. Once again, scroll the animation to the left and right to compare the section from Taxon 1 with the sequence of bases identified for Taxon 2. When you have found the highest match you can, click on the 'sequences aligned' button and check your answer.

Other reasons for a mismatch between sequences can be through addition, removal, inversion or duplication of a base or series of bases. Screen 4/8 shows two more sequences of nucleotide bases, but this time we want to identify what the 'copying error' was that introduced the mismatch. First, find the best match between sequences, shown in Text-fig. 16.

Note that there are two sizeable lengths which match perfectly, the five bases on the far left (AGGAT) and the three on the far right (CCC). An A matches in the middle, but notice that in both taxa this is part of an AATCCC sequence, which is displaced relative to its counterpart in Taxon 2, as shown in Text-fig. 17.

It looks as if Taxon 2 has lost the G immediately before the AATC-CC sequence, and that that sequence has been 'shunted' to the left. If you click on the 'deletion' option this answer is confirmed. The animation allows you to move the Taxon 2 sequence to the left and the right. If you do this by a single base in either direction you can see that the AGGAT and AATCCC sections do indeed match in alternation.

Just as homologies can be found in molecular data, so can homoplasy. With successive generations, 'shuffling' in the genome brings about similarities which do not imply homology. Similarities between sequences can come about through convergence and through reversals, as seen in

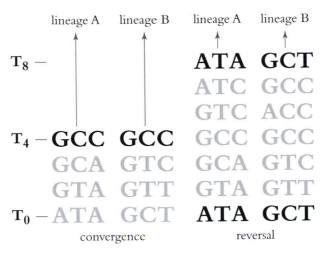

Text-fig. 18

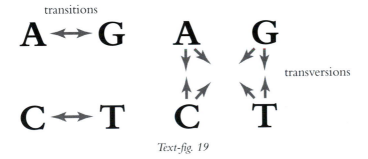

Text-fig. 19

Text-fig. 18. On the left-hand side are two lineages plotted for the first four time intervals (T_0 to T_4). Notice that although the three bases began differently (ATA for lineage A and GCT for lineage B) they are now the same, GCC. This is a *convergence*, since this similarity does not imply a common ancestor. On the right-hand side the two lineages are plotted for a further four time intervals. At time T_8 the two lineages now have the same three bases as they did initially. This is a *reversal*, because it was not an absence of changes from time T_0, but rather a series of changes 'there and back again' that resulted in the same three bases.

Although any one base can be replaced with any of the other three (Text-fig. 19), it is much more common for one pyrimidine to be switched for the other, and purines likewise, i.e., for adenine (A) to be switched for guanine (G) and cytosine (C) for thymine (T). Such exchanges are called

31

transitions. Transversions, or the replacement of a purine with one of the two pyrimidines (or a pyrimidine with one of the two purines), are much less common.

On Screen 7/8 we are presented with a three-lineage 'tree' based on six nucleotide bases. We are asked to determine which bases are apomorphic, i.e., derived character states, at time interval 10 for (a) the left and centre lineages and (b) the centre and right lineages. Let us look at each character in turn.

1. The plesiomorphic (or primitive) state for this base is A, and that state is observed in the left and centre lineages as well. Only the right-hand lineage has a derived state for this character, having switched to a T at time interval 5 and a C at time interval 7.
2. The plesiomorphic state is C, but all three lineages show the derived state T.
3. The plesiomorphic state is T, but only the right-hand lineage has a derived state for this character, C.
4. The plesiomorphic state is G. The left and centre lineages have the derived state A, while the right lineage retains the plesiomorphic state G.
5. The plesiomorphic state is T. The left and centre lineages both have the derived states, but different ones – G on the left and C in the centre. The right lineage retains the plesiomorphic state T.
6. The plesiomorphic state is A. The left and centre lineages have the derived state G, while the right lineage retains the plesiomorphic state A.

Base 2 is not informative as all three share the same derived state, so cannot be apomorphic for either pair of lineages exclusively. Bases 1 and 3 are derived in only one lineage, the right-hand one, and so are also un-informative. Likewise, base 5 is uninformative; although the left and centre lineages show derived states, they are different ones, and so cannot be apomorphies. However, bases 4 and 6 are informative, being apomorphic for the left and centre lineages. So to answer this question correctly, click boxes 4 and 6 for row (a).

Notice that while these data suggest a close relationship between the two lineages, this is misleading. The actual phylogeny shows that the right and centre lineages diverged more recently (between time intervals 4 and 5) than the left and centre lineages (between time intervals 0 and 1). What has happened is that base changes have created apparent, but false, synapo-morphies which obscure the true phylogeny. This is a real problem for taxonomists. The best solution is to choose long sequences of bases, and preferably ones that are relatively stable over successive generations.

2.3 – Character definition
Screens 1/23 to 7/23

As we have seen, characters are fundamental to investigating the relation-ships between groups of organisms, but the trick is in deciding which ones are informative (i.e., the *homologies*), and which ones are either unin-formative (such as *plesiomorphies*) or downright misleading (i.e., *homo-plasies*). On Screen 2/23 we are presented with aligned base sequences from four different taxa. Some of these bases might be useful … but which ones?

The very first site is uninformative because the same base pair occurs at this site for all four taxa; it is what is known as an *invariant site*. In this analysis, this particular character does not discriminate one group of taxa from the others. In fact, there are plenty of other invariant sites, as shown with arrows in Text-fig. 20.

The fourth and fifteenth sites are also uninformative, although for a different reason. In these cases, only one taxon has a different base to the others at these particular locations. Since the acquisition of a unique char-acter state by one taxon is compatible with any of the possible networks of relationships, such a character tells us nothing about the relationships of the four taxa. Instead of being invariant, the bases at these sites occu-py *phylogenetically neutral sites*; that is, while demonstrating some change, they do not favour any one network of relationships over another. On Text-fig. 20 these have been marked with asterisks.

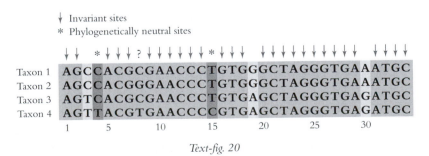

Text-fig. 20

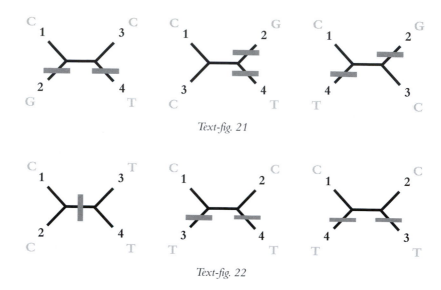

Text-fig. 21

Text-fig. 22

Is site 8 (with question mark) also phylogenetically neutral? Taxa 2 and 4 both have different but unique bases at that point while 1 and 3 share the same one. Let us draw out the three possible networks of relationships (Text-fig. 21).

Clearly, two state changes are needed to make these data fit any of the networks, so this character is phylogenetically neutral, and therefore uninformative. Sites 3, 19 and 30 are potentially informative, though. The bases at these locations are neither all the same nor different in only one taxon. At each of these sites two taxa share one character state while the two other taxa share another state (e.g., C or T at site 3). Mapped out against the three possible networks of relationships, one tree is more parsimonious than the other, as shown in Text-fig. 22.

Characters such as these could reveal something of the phylogeny of the four taxa by identifying groups with shared, derived character states, i.e., synapomorphies. In Text-fig. 20 these have been highlighted in light grey.

On Screen 6/23, a fifth taxon is added to the data matrix. This turns site 8 from an uninformative, neutral site to an informative site, because now two shared character states for two groups of taxa are revealed (C by Taxa 1 and 3, and G for 2 and 5). So, invariant sites, phylogenetically neutral sites, and informative sites are not absolute qualities but rather variable ones, depending on the level of the analysis. Adding or removing taxa can make useful characters uninformative, and vice versa.

Screens 8/23 to 11/23

One of the simplest ways to define a character is on its presence or absence; for example, the presence of hair on the body is a character that can distinguish mammals (which have hair) from all other vertebrates (which do not). Other examples include antlers on deer, feathers in birds, the swim bladder in bony fish, and the phloem and xylem in plants. Such characters are called *binary characters*; in a character matrix their presence is often denoted by the value 1 and their absence by a 0.

Because binary characters are so unambiguous, they are the most convenient characters to work with, and biologists have often used them to define groups of organisms. However, there are many other characters that cannot be defined simply on their presence or absence, but must be defined instead on some variable attribute. Many mammals have different numbers of toes, for example (see Screen 10/23). These sorts of characters are called *multistate characters*. Almost anything that occurs in different numbers in some aspect of the morphology of an organism might be a multistate character, for example the number of spines on the dorsal fin of a fish, or the number of petals on a flower.

The problem with multistate characters is that, unlike binary characters, there is no single, logical and obvious way of defining the various states and entering them into a character matrix. A binary character is an 'either/or' statement, but defining of the number of toes can be done in many ways, all of them equally valid, though not necessarily equally useful. You could divide it up into five states, one for each number of toes, and assign them values from 1 to 5. So a horse, with one toe, would have the value 1, a cow, with two toes, the value 2, and so on up to humans with five toes, a value 5. Alternatively you could do the same thing but numbered backwards, or even in no particular order. Yet again you could divide it into only two states, even numbers of toes and odd numbers of toes, or perhaps into prime numbers and non-prime numbers. Actually, all that matters in practical terms is that you are consistent and explicit, so that any other taxonomist can recognise the character states you have defined (even if they would do it themselves in a different way!).

Molecular data can be handled either as multistate characters or as binary characters. As multistate characters each of the bases is counted as one state, while as binary characters pyrimidines count as one state and purines as the other.

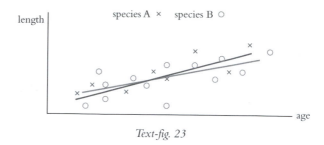

length   species A ×   species B ○

age

Text-fig. 23

Screens 12/23 to 14/23

While multistate characters can be used to describe many variable characters, yet more cannot be broken down into discrete categories. A bony fish will have a certain number of spines in its dorsal fin, and that number will be an integer, and so discrete from another character state having one extra spine or one fewer spine. But many characters have states which grade into one another. Characters based on dimensions of some sort are often of this kind, for example length or the ratio between width and height.

Imagine a character based on length taken from a population of organisms which are suspected to include two different species, for example the total lengths of a number of Lake Victoria cichlid fishes. When the length of individuals of different ages is plotted, a line chart such as that shown in Text-fig. 23 is produced.

There is no obvious pattern to these data, and so no easy way of dividing the lengths of these fishes into the small number of discrete character states that a character matrix needs. The best that can be done is to divide the length character into arbitrary units. Screen 13/23 shows another example of this sort of character, the relative size of the middle toe compared with the other toes in a lineage of fossil and living horses.

Compare this with a similar plot for a different population, say of the shoulder heights of wild asses and wild horses against age, as shown in Text-fig. 24. There are two obvious 'clumps' of heights, suggesting that height could be divided into two distinct characters, with the lower range of heights being character state 0 and the higher range of heights character state 1.

The example given on Screen 14/23 is the ratio of width and length of various Galapagos island finches. Again there are three distinct clumps

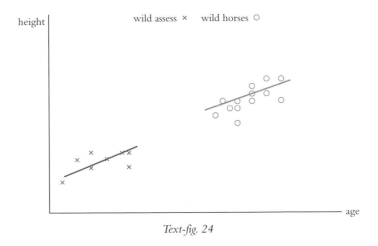

height | wild assess ×   wild horses ○

age

Text-fig. 24

of ratios, one for each species, and so if this character were being entered into a character matrix these would be the three character states to define.

Screens 15/23 to 23/23

This section is a series of exercises testing what we have learned about defining characters and character states. The organisms in question are trilobites, extinct, woodlouse-like marine invertebrates that were common during the Palaeozoic era. Being arthropods, they had jointed external skeletons which they periodically cast off as they grew, and these moults are particularly abundant in the fossil record. If you move the cursor over the four small photographs of the different trilobite species in Screen 16/23, you will see a magnified view in the centre of the screen. For each one, try to identify the parts as labelled in Text-fig. 25.

Trilobite structure can be divided into the cephalon, the thorax and the pygidium. The head is the large semi-circular section at the front. On top are two compound eyes and the central glabella, which housed some of the digestive apparatus. The thorax is made up of a number of segments, each of which bore a pair of biramous (or two-armed) appendages. The top appendage was a gill and the lower appendage a leg. At the end of the animal was the pygidium, which was made up of a number of segments rather like those of the thorax, but more or less fused into a single shield. When the trilobite was disturbed it could roll up like a pillbug, with the pygidium and cephalon held tightly together.

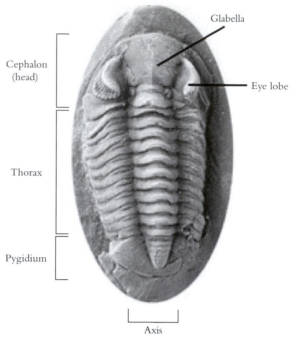

Cephalon
(head)

Thorax

Pygidium

Glabella

Eye lobe

Axis

Text-fig. 25

Screen 17/23 asks what sort of character the number of segments in the thorax would be. Since there is a certain (integer) number of them, such a character can be divided into a number of discrete states. Thus this character is a multistate character. The next screen (18/23) asks what sort of character the proportional length of the glabella relative to the cephalon (i.e., the ratio between the two lengths) might be. As these two lengths vary continuously, this is going to be a continuously variable character. On Screen 19/23 the same question is asked about the presence of eye lobes. As these are either present or absent, this is going to be a binary character. Next we are asked about the 'glabellar furrows', the indentations on the glabella seen in some species (Screen 20/23). As there are a certain countable number of them, th is must be a multistate discrete character. Screen 21/23 describes a character based upon a comparison of two continuously variable dimensions, the width of the axis and the width of the thorax; this must be a continuously variable character. The next screen (22/23) asks about the presence of conspicuous reticulation on the cephalon. Reticulation is a style of ornamentation resembling a fine net, and as far as we can tell is present in one species

(*Aulacopleura*) and absent in the others, so this must be a binary character. Finally, the ribbing on the pygidium, formed by the fusion of the last few segments, is a countable number, more in some species and fewer in others. Such a character is going to be a multistate character.

2.4 – Weighting

Screens 1/3 to 3/3

Weighting characters is a very controversial aspect of phylogenetic analysis. In essence, advocates of character weighting argue that some characters have more 'worth' than others in deducing the pattern of relationships between organisms. Intuitively this seems reasonable enough, but the problem is that determining which characters are more valuable than others, and so what relative weighting, or influence on the analysis, they should be assigned, is a subjective process. With all the problems of homoplasy, reversals and convergence that we have already seen, it is not difficult to imagine a situation where characters that appear to be important in fact are not, and so by weighting them early on in the analysis they obscure rather than reveal the true phylogeny.

Opponents of such *a priori* weighting prefer to give all characters equal value, at least in the initial stages of the analysis, and allow the greatest degree of congruence between the largest number of characters to separate useful characters from misleading ones. We saw an example of this earlier in this section when comparing the congruence among vertebrate eyes and limbs (homologous characters) with the convergent but homoplasious appearance of wings in birds and insects.

Once the parsimony analysis has been performed, some taxonomists advocate re-weighting characters relative to the amount of homoplasy they show. Characters which are characteristic of a certain clade only are clearly showing a strong 'signal', and can be weighted upwards relative to the other characters. Characters which show no consistency but instead show lots of reversals and convergences on different parts of the tree instead generate 'noise'. These can be weighted downwards. If the analysis is run again the noisy characters will have a weaker effect on the topology of the tree than the ones with a strong signal, and it is hoped the tree will be all the better for it. This is called *a posteriori* weighting.

3–Cladograms and trees

3.1 – Rooting procedures and character polarity
Screens 1/11 to 5/11

The direction in which character state changes are believed to have occurred over the evolutionary history of a lineage is called the *character polarity*. On Screen 1/11 the cladogram that was deduced earlier for the four insects (the butterfly, moth, house fly and crane fly) is shown again. There are two distinct clades within the four taxa, one comprising the moth and the butterfly, and a second containing the house fly and the crane fly. The moth and butterfly clade can be defined by a synapomorphy, the presence of scales on the wings, a feature which the house fly and crane fly lack. Likewise, the house fly and crane fly clade also has a synapomorphy, the presence of halteres instead of a second pair of wings. In that example we were told that scaly wings and halteres represented derived states, i.e., that the ancestors of all four species had plain, not scaly, wings, and that two pairs of wings was the primitive state and halteres were derived from the second pair of wings. But how can such character polarity be determined?

There are two main ways. One is by using the *ontogenetic criterion*, i.e., comparing the way the character develops in the species being studied with the occurrence and development of that character in its putative ancestors. The other way to establish the polarity of a character is to make *outgroup comparisons*.

Consider swim bladders in the teleosts, or bony fishes. In some fishes, such as the garpike (*Lepisosteus*) and bowfins (*Amia*), the swim bladder is highly vascularised and connected to the gut through a pneumatic duct.

These fish periodically gulp air, which is passed into the swim bladder and from which oxygen is extracted. Functionally, the swim bladder is a lung. Other species, such as carp and catfishes, also have a swim bladder which is connected to the gut, but it is not highly vascularised. These fish do not use the swim bladder as a lung but instead use it as a buoyancy organ only; they still periodically gulp air to inflate the swim bladder. Yet other kinds of teleost, such as perch and bass, have the pneumatic duct only as juveniles; in adults there is no connection between the swim bladder and the gut, and so they cannot gulp air. Instead they use a special 'gas gland' to inflate the swim bladder using gases from the blood stream. This allows them to live in deep water habitats, where coming up to the surface to gulp air would be impossible.

Using a variety of morphological and molecular characters, the phylogeny of bony fishes has been determined to be that shown in Text-fig 26.

From this it can be observed that the most inclusive group (which includes all six species listed here), the entire teleost clade, shared an ancestor having a lung-like swim bladder. Two of the species listed here (the 'outgroup' species) retain this condition. Within the teleost clade are fishes which have lost the vascularised swim bladder (the 'intermediate' and 'advanced' species), and within that clade is another clade which has lost the connection between the swim bladder and the gut (the 'advanced' species).

Looking at the way the advanced fishes develop, the connection between the swim bladder and the gut reflects the presumed evolution of these fishes from more primitive ones. The juveniles of the advanced species have the connection, but it is lost in adults, so in this respect there is a morphological

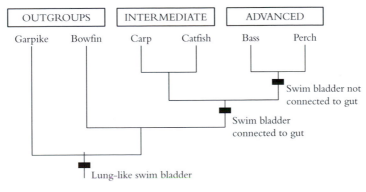

Text-fig. 26

similarity between juvenile advanced fishes and juvenile *and* adult intermediate fishes. This is an example of Von Baer's Law, that the general (plesiomorphic) features appear in the development of an animal before the more derived (apomorphic) ones. The connection of the swim bladder to the gut is common to the more inclusive grouping, the intermediate + advanced fish clade, while the loss of the connection is a characteristic of the derived grouping, the advanced fish clade. Observations like these can help biologists to establish the polarity of a character, i.e., in which direction character states advanced within the evolution of a clade.

Screen 2/11 presents a group of sand dollars (burrowing sea-urchins). Certain species have openings in their shells called lunules. If the ontogeny of one such species is examined (Screen 3/11) it can be observed that juveniles lack lunules, and that the lunules develop as the animal matures. By reference to the ontogenetic criterion, then, lunules would appear to be derived characters, and so absent from the ancestors of the clade of sand dollars which possess them (which includes *Leodia* and *Echinodiscus*). The alternative approach is outgroup comparison, as shown on Screen 4/11. Comparing the clade including *Leodia* and *Echinodiscus* with the other two sand dollars, as well as other sea-urchins, again suggests that lunules are a shared, derived feature of the sand dollars which have them, i.e., a synapomorphy.

Screens 6/11 to 11/11

The exercise presented in this series of screens is a demonstration of how determining the polarity of characters is used in a parsimony analysis, in this case a study of the phylogeny of a particular genus of pocket gophers, *Orthogeomys*. (Gophers are a kind of burrowing rodent.) The characters used happen to be molecular ones, but the notion of polarity as described earlier can still be applied. Of course, ontogenetic criteria are not relevant here – an organism's genome remains constant throughout its life – but comparisons with outgroups (more distant relatives) can be made. Another species of gopher *Thomomys talpoides*, is given as an outgroup.

The first step is to identify the informative characters from the character matrix shown on Screen 7/11. Remember, these molecular characters can be divided into informative sites, phylogenetically neutral sites and invariant sites. The invariant sites are the easiest to spot. In these, the ingroup taxa (i.e., the four species of *Orthogeomys*) share the same character state (A, C, G or T) as the outgroup taxon, *Thomomys talpoides*. This can be observed to be the case for character 1, for example, where all five taxa have the base T. The other invariant sites are characters 5, 7 to 11, 13, 16, 17, 19, 20, 23 to 26, 29 to 31, 33 to 37 and 39 to 43. The phylogenetically neutral sites exhibit a character state which, although at least one ingroup taxon has a different state to that of the outgroup, does not favour any one arrangement of the four ingroup species as being most parsimonious. This is the case where only one ingroup taxon is different, for example with character 2. There one species has the base G while the other three species share the same state as the outgroup, namely base A. The alternative phylogenetically neutral condition is where all four ingroup species have a different state to the outgroup, as is the case with character 18, where all four *Orthogeomys* have the base T whereas the outgroup has the base G at this position. Besides characters 2 and 18, the phylogenetically neutral characters in this matrix are 3, 6, 15, 21, 23, 27, 32 and 38. Only the remaining five characters are phylo-genetically informative: characters 4, 12, 14, 28 and 44.

The next step is to plot these characters and the character state changes involved against all the possible cladograms to determine which cladogram is the most parsimonious. Since the position of the single outgroup species is fixed by definition to the bottom of the cladogram, the possible cladograms will differ in the position of the four ingroup species, and as we saw in Section 1.2 there are fifteen ways to arrange four taxa. These fifteen trees are presented on Screen 8/11, and you can either print them off or draw them yourself. Text-figure 27 is one such tree (on the display of the trees on Screen 9/11 it is the first tree). It has been labelled with all the character state changes. A similar sort of thing can be done for any of the possible trees.

In this particular case, there are eight character state changes in total, and so the cladogram is said to have a length of eight steps. The other possible cladograms will have more or fewer steps depending on their topology – the one with the fewest steps, the 'shortest tree', is the most parsimonious. In fact, of all the trees it is Tree 7 that comes out as being the most parsimonious, as shown in Text-fig. 28.

Since there are a total of five character state changes, this tree has a length of five. It is shorter than all the other trees, which vary in length from seven to ten. By the principle of parsimony, then, this is the most likely phylogeny for the genus *Orthogeomys*.

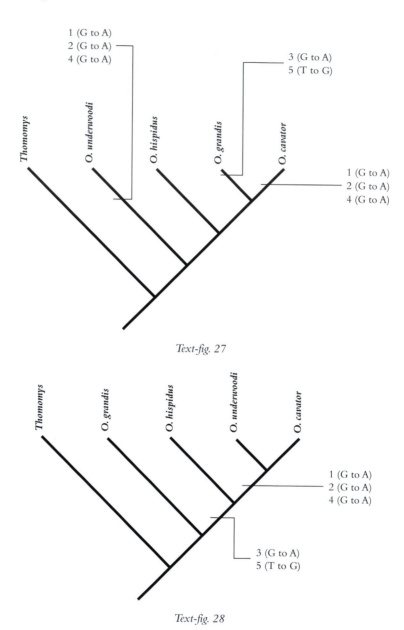

Text-fig. 27

Text-fig. 28

In real life, biologists rarely draw out all the possible trees and compare them one with another. With more species and more characters the number of possible trees soon becomes impossibly large. Instead, computers are used to analyse data sets. Computer programs exist which contain algorithms that hunt out the most parsimonious trees possible for any given data matrix. Examples of such programs are PAUP and HENNIG (see the Appendix). Since large data sets can produce millions of possible trees, it is not uncommon for biologists to set aside their computers for hours or even days at a time to run the analysis.

3.2 – Cladograms, phylograms and phylogenetic trees

Screens 1/10 to 10/10

Cladograms, phylograms and phylogenetic trees are all related, being different methods of portraying the same data in such a way as to describe a hypothesis of relationships between taxa. However, they do differ in significant ways (see Text–fig. 29).

A cladogram is a schematic representation of the hierarchy of relationships between taxa. The lengths of each branch are purely arbitrary, and say nothing about the numbers of character state changes on each branch, or when the branching events happened. On a cladogram, the important aspect of the tree is its topology. The nodes, or junctions between branches, do not specifically relate to any known or unknown taxon, so shouldn't be thought of as 'ancestors' to the lineages which diverge from them. They denote only the relative recency of common ancestry.

A phylogram is similar to a cladogram, again having a topology which reflects the hierarchy of relationships. However, instead of arbitrary branch lengths, each branch length is proportional to the number of character state changes that define it. In Text–fig. 29, the branch leading to Taxon B is 'one step long', i.e., only a single character state change occurs. The same is true for the branch leading to the clade comprising both Taxon B and Taxon C. On the phylogram, the number of character state changes on a branch is marked by horizontal bars, so the branch leading to Taxon A is three steps long, the one leading to Taxon C four steps long, and so on.

Finally, a phylogenetic tree again shares the same topology as the cladogram, and therefore the presentation of the hierarchical relationships between the taxa, but adds time data along one axis. The stratigraphical ranges of the various taxa are plotted and connected with a pattern of branches compatible with the topology of the cladogram. In the example in Text–fig. 29, Taxon A ranges from 50 to 40 million years ago, Taxon B from 40 to 30 million years ago, and Taxon C from 10 million years ago to the present. Note that the stratigraphical ranges are not the same as the implied branches between the taxa. Stratigraphical ranges are deduced

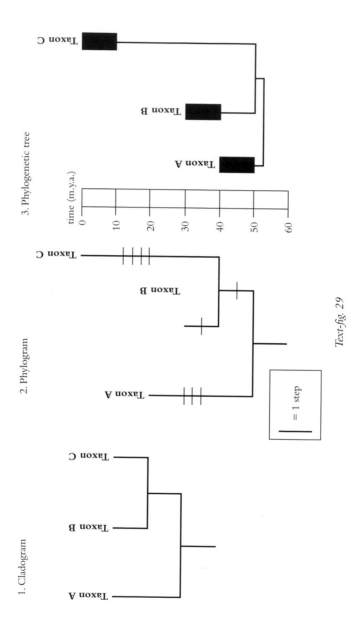

3. Phylogenetic tree

time (m.y.a.)

Taxon C

Taxon B

Taxon A

0
10
20
30
40
50
60

2. Phylogram

Taxon C

Taxon B

Taxon A

= 1 step

1. Cladogram

Taxon C

Taxon B

Taxon A

Text-fig. 29

from the first and last appearances of a taxon in the fossil record – but there are plenty of reasons why the fossil record for any given taxon could be incomplete. Many organisms have a very patchy fossil record, with only odd fragments being found occasionally, as is the case with dinosaurs and fossil hominids, for example. So the fact that Taxon C diverged from Taxon B no later than 50 million years ago, but yet is not known from any fossils older than 10 million years, is just the sort of problem palaeontologists encounter when looking at real phylogenies. Such implied existences of taxa even where the fossils are lacking are called 'ghost lineages'. One of the best-known cases where an extensive ghost lineage is implied is the modern coelacanth *Latimeria chalumnae*. The youngest fossil coelacanths date from the Late Cretaceous, about 70 million years ago, and coelacanths have not yet been found in rocks of Tertiary or Quaternary age. But the very fact that coelacanths are alive and well in today's oceans means that there must have been coelacanths during that intervening period of time.

The screens that make up the rest of this section show a real example of such different kinds of trees, using a group of bivalve molluscs called rudists, which were particularly common during the Cretaceous. Screen 3/10 shows the cladogram for the species being studied. Notice that not every taxon is defined by a character transformation. *Pachytraga tubiconcha*, for example, is part of a clade including three other species (*Pachytraga paradoxa*, *Offneria simplex* and *Praecaparina varians*) which as a group share the transformation of character 3 from 0 to 1. However, *Pachytraga tubiconcha* can be separated from the three other species in its lacking the transformation of character 7 from 0 to 1, hence its position on a separate branch of the cladogram. But with respect to the characters listed in the data matrix here, it does not have any unique, derived characters of its own.

The phylogram (Screen 5/10) shows this clearly. Here, branch length is proportional to the number of character transformations, and those species that are not defined by any such character transformations are on the end of 'zero-length' branches. Taxa such as these could have been simply outliers to the main line of evolution within the clade, retaining the original defining features of the group but not adding any of the later features or new ones of their own. Sometimes they can even be considered to be not just primitive with respect to the more derived members of the clade but actually ancestral to all the others. However, there are good reasons why many palaeontologists shy away from describing species as ancestors, not the least of which is the incompleteness of the fossil record. While a species

such as *Pachytraga tubiconcha* might not have any apomorphic shell characteristics, we cannot say if the same is true for characteristics of its bio-
chemistry, behaviour or genes. Perhaps it had some aspects of its soft body
anatomy that would have provided useful apomorphies, but since the fossil
record doesn't retain them we cannot tell.

Finally, Screen 9/10 plots the cladogram against time to produce a
phylogenetic tree. There are some ghost lineages implied by the phylogeny, but what is important on a practical level is that the order in
which the species occur broadly matches the phylogeny, i.e., the basal
species are from the older sediments, and the more derived species from
younger sediments. To many palaeontologists, this is a more important
test of a phylogeny than whether or not ghost lineages are needed to
make the cladogram 'fit'.

3.3 – Monophyly, paraphyly and polyphyly

Screens 1/14 to 8/14

Cladograms are used to describe the relationships between organisms
with respect to shared, derived features – synapomorphies. Where traditional taxonomic groupings, such as the Mammalia, can also be defined
on the basis of synapomorphies, we can say with confidence that such
groups evolved from a common ancestor and all share certain derived
features (Screen 1/14). Such a group is said to be *monophyletic*, i.e., a clade.
Another monophyletic group shown on this cladogram (Screen 2/14) is
that of the vertebrates (or Vertebrata), which share the development of a
backbone. A third is the clade of triploblastic animals (Screen 3/14), all of
which have three fundamental layers of tissues – the endoderm, the
mesoderm and the ectoderm – from which their bodies develop.

On the other hand, the invertebrates are a group defined as animals
without a backbone (Screen 5/14). Since a backbone is a synapomorphy
of the vertebrates, the fact that invertebrates lack one simply reflects their
exclusion from the vertebrates. The absence of a backbone is therefore a
shared, primitive feature of the invertebrates, i.e., a *symplesiomorphy*.
Symplesiomorphies cannot be used to define clades. Instead, the invertebrates could be described as what's left over when one clade (the
vertebrates) is removed from another (the animals, or Animalia). Such
'left-overs' are called *paraphyletic groups*. The Reptilia, as traditionally

defined to include snakes and alligators, are another paraphyletic group, as can be seen on this cladogram. Snakes, alligators and birds form a clade, as do the birds on their own, but there is no clade consisting of the snake and the alligator alone. Indeed, cladistic analysis and recently discovered fossils of feathered dinosaurs are making it increasingly clear that birds and what we call reptiles are difficult to separate. Although a paraphyletic group like the reptiles is not a clade, it still conveys some phylogenetic information, including a series of taxa increasingly closely related to the ingroup clade. Within the reptiles, for example, the alligators can be (perhaps surprisingly!) described as the closest living relatives to the birds.

The radial body plan of corals and starfish, as opposed to the bilateral symmetry of snails, crabs and birds, could be used to divide up these animals into two groups. But such groups would not be clades. While superficially similar, corals and starfishes have many fundamental differences in structure, development and physiology. Indeed, the starfish and the coral lie in quite different parts of the tree. A grouping of the coral and the starfish would therefore be neither a clade nor even a paraphyletic grouping. It is a *polyphyletic group*, one which 'cherry picks' taxa from widely different parts of a cladogram to make a group based on some superficial, but cladistically misleading, character. Unsurprisingly, polyphyletic groups are not phylogenetically informative.

Although they have no place in a strictly cladistic classification, both paraphyletic and polyphyletic groups occur in traditional taxonomies. Indeed, many have been created as a result of the *ad hoc* way that traditional taxonomic groups have been defined, using whatever apparent characteristics seem important at the time, without reference to phylogeny. While some biologists maintain that paraphyletic groups can have some usefulness, none would defend the use of polyphyletic groups.

Where a pair of clades diverge, for example the starfish on the one hand and the vertebrates on the other, biologists often refer to the two clades as *sister groups*. The two sister groups themselves make up a bigger clade, in the case of the starfish and the vertebrates one called the Deuterostomes.

Screens 9/14 to 14/14

These screens test your understanding of the concepts of monophyly, paraphyly and polyphyly. Look closely at the tree before answering each question. Try to determine whether any given group of taxa is a clade or not. Remember, by definition, a monophyletic group must be a clade. Sister groups are also clades, which together form a larger clade. A paraphyletic group is what's left over from a larger clade when a smaller clade is removed from it, while a polyphyletic group is simply a group of taxa which is neither a clade nor a paraphyletic group.

On Screen 9/14, the alligator and the snake form a paraphyletic group relative to the birds, but being paraphyletic cannot be called the sister group of the birds. Likewise, the triploblastic invertebrates are a paraphyletic grouping, and so cannot be a sister group to anything else (Screen 10/14). In contrast the diapsids (including the birds, alligators and snakes) are a clade, and one which is the sister group to the clade of mammals (Screen 11/14). A grouping of jointed limbed animals (including the crab, alligator, bird, panda and human) is clearly polyphyletic, leaving out lots of intervening members of the clade from which they are drawn, i.e., the worm, snail, starfish and snake. Not being a clade, it cannot be a sister group to anything else (Screen 12/14). Finally, a group including the crab, worm and snail, known as the Protostomes, is a clade (Screen 13/14), and one which is the sister group of the Deuterostomes.

3.4 – Consensus trees
Screens 1/11 to 11/11

As mentioned in Section 3.1, real-life data sets can produce thousands or millions of trees. If you are lucky, there is only a single most parsimonious tree, but often there are more than one. If the number of such trees is small, and you have some suitable criteria to select one from the others, then it is possible to choose one of these most parsimonious trees as the one you prefer. For example, if you are working with fossil organisms, you might choose the one tree that best matches the stratigraphical ranges of the taxa, i.e., in terms of minimising conjecture about inferred but unrecorded stratigraphical distributions, is the most parsimonious. But if there are simply too many trees to inspect and no reasonable criteria to discriminate between them, it is impossible to choose one tree as the 'best'. Instead, biologists determine consensus trees, which can be thought of as averaging out all the trees so as to highlight the least contentious topological features. Of course, it is essential to realise that like many statistical averages they have no reality in and of themselves – in the same way that the average shoe size of a group of people might be 6.34 but not one

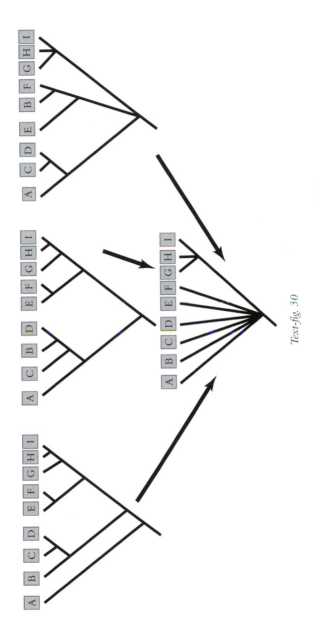

Text-fig. 30

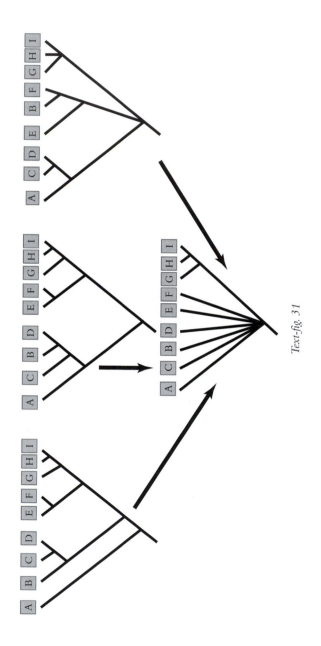

Text-fig. 31

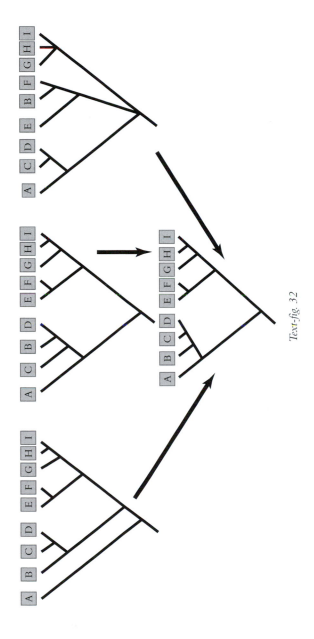

Text-fig. 32

of them has shoes that size! The consensus trees simply reflect common features among all the most parsimonious trees. Most importantly, they are not phylogenies, even though they look like cladograms.

The *strict consensus tree* is the simplest sort of consensus tree, containing only those branches that are found in all the most parsimonious clado-grams (Screen 2/11). Nodes that are not shared by all the cladograms are collapsed. Although collapsing nodes not shared by all the trees gets rid of some data, the advantage of strict consensus trees is that they highlight the ubiquitous features, and so are the least statistically processed kind of consensus tree. This is shown clearly on Screens 3/11 and 4/11. Almost all of the topology of the three trees is absent from the strict consensus tree, but the latter does show that there is some sort of consistent clump-ing of Taxa G, H and I relative to the other taxa (Text-fig. 30).

An alternative to the strict consensus tree is the *semi-strict consensus tree*. The semi-strict consensus tree contains both nodes ubiquitous to all the trees and any nodes on one or more trees not contradicted by nodes on the other trees (Text-fig. 31).

Compared with the strict consensus tree, the semi-strict consensus tree is a bit more resolved. While the polytomy of Taxa A to F remains, the topology of the clade including Taxa G, H and I is fully resolved. (A polytomy is a node in a cladogram bearing multiple branches.) On the negative side of the balance, it must be understood that the consensus tree now contains some arrangements of taxa not found in one of the trees, specifically the arrangement of G, H and I. In other words the semi-strict consensus is introducing a statistical artefact that may not be present in the actual phylogeny.

A third type of consensus tree transmits even more of the topological data common to at least a large proportion of the trees, and that is the *majority rule consensus tree* (Text-fig. 32).

The majority rule consensus tree contains any nodes which occur in more than an arbitrary percentage of the trees, usually 50% but it can be a higher value, and the percentage of trees sharing each node is often plotted on the diagram in some way. This is a very widely used technique for when strict or semi-strict consensus trees have too little resolution to allow even the most basic interpretation of the results of a cladistic analy-sis. The higher the percentage of trees which share any given node, the stronger the support for that particular node is considered to be.

4–Fit and robustness

4.1 – Measuring goodness of fit

Screens 1/33 and 2/33

Having found the most parsimonious tree (or trees) for any given data set, the next problem facing a biologist is deciding which nodes reflect real phylogenetic patterns, and which ones have come about through some homoplasious process such as convergence. In short, one needs to be able to differentiate the signal from the noise. There are essentially two ways to do this. First, one can look to see how well the various characters fit on the most parsimonious trees. Characters which undergo similar transformations at many different points on the tree are clearly homoplasious, while characters that change only once and thus define clades are not homoplasious. Establishing the consistency of a character on a cladogram is known as measuring the *goodness of fit* to the tree. The second way to draw out the phylogenetic signal from the background noise on a cladogram is to test individual nodes rather than the characters, and see just how *robust* they are. A strongly supported node will be less affected by errors in the data matrix than a weakly supported one. A robust node reflects not just a particular pattern in the data but probably also a real historical event, and so 'messing up' the data slightly shouldn't collapse that node. By contrast, a weak node may be simply a chance correspondence of characters and taxa, and a slight change to the data matrix will destroy it. Section 4.1 looks at how characters can be tested for homoplasy, while Section 4.2 discusses testing the topology of the cladogram itself to measure the robustness of the nodes.

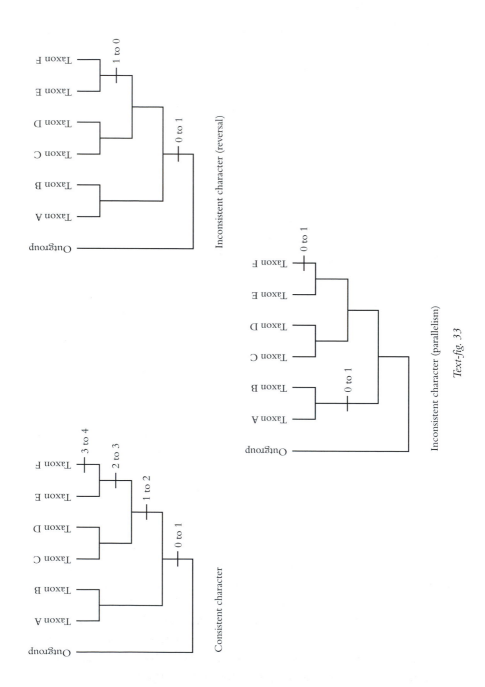

Text-fig. 33

Screens 3/33 to 11/33

Tree length is simply the number of character transformations that the tree contains. The most parsimonious tree (or trees) is the one with the fewest transformations, and is said to have the shortest length (or to be the shortest tree). Other, less parsimonious trees will have more character transformations and so are trees of greater length. Assuming that the most parsimonious tree is the one that is most likely to reflect the true phylogeny of the group of organisms being studied, it is possible to look at how often each character underwent a transformation from one state to another. Characters which are likely to provide useful phylogenetic information will undergo transformations only once or a few times on the tree, without reversals or similar transformations on different parts of the tree (parallelisms). By contrast, homoplasious characters show repeated similar transformations on many different parts of the tree (Text–fig. 33).

Screens 4/33 to 10/33 give examples of how the character states present in the data matrix can be compared with the character transformations of the cladogram. The sum of all of these transformations is the tree length. Screen 11/33 tests your ability to determine tree length and so decide which of a number of competing trees is the most parsimonious.

Screens 12/33 to 20/33

It is important to understand that the most parsimonious tree will also be the one with the least inferred homoplasy. Homoplasy can be defined as the extra number of transformations, or steps, a given character undergoes on a cladogram compared with the theoretical minimum number of steps that the character could undergo to exhibit all its character states. For example, a character with two possible states (0 and 1) has a theoretical minimum number of transformations of one, i.e., from 0 to 1, if 0 is the primitive state, which we cannot assume *a priori*. Any additional steps above this minimum number of steps imply that the character is displaying some degree of homoplasy; the more additional transformations, the greater the amount of homoplasy. But if that character undergoes only a single transformation then the cladogram is displaying the theoretical minimum number of steps for that character, and so the character is showing no homoplasy.

There is a simple index for quantifying this variance between the theoretical minimum number of steps and the actual number of steps for that character which can be seen on a cladogram – the *consistency index*, or *ci*. The consistency index is the theoretical minimum number of steps for a given character divided by the number of steps it displays on the cladogram, and has a value between 0 and 1 (Screens 13/33 to 17/33):

$$\text{Consistency index} = \frac{\text{Theoretical minimum number of steps}}{\text{Observed number of steps}}.$$

The lowest values occur when there are many more observed steps for the character as seen on the cladogram than the theoretical minimum, and so imply a high degree of homoplasy. When the observed and theoretical minimum number of steps are equal, then $ci = 1$, which means that this character is not homoplasious. So, the higher the ci value, the less homoplasy is inferred for the character to fit the tree, and so the more useful the character is likely to be in establishing the phylogeny of a group of organisms.

A consistency index can also be calculated for the entire tree (Screen 17/33). By dividing the sum of all the theoretical minimum number of steps by the sum of all the observed number of steps, another value between 0 and 1 is obtained. This is the *ensemble consistency index* (sometimes simply referred to as a tree's consistency index) or *CI*. As with individual characters, the higher the value, the lower the amount of homoplasy inferred to be present in the tree. While the ensemble consistency index can be useful in helping to choose between a number of equally parsimonious trees from a single data set, it is not to be used to compare trees from different data sets. Bigger data sets tend to have lower consistency indices than smaller data sets, because there is a negative correlation between the size of the data matrix and the consistency index.

Screen 21/33

The consistency index, however, fails to take into account where on a cladogram a character transformation occurs. As such, it cannot distinguish between a character that changes at the base of a clade – a synapomorphy – and one that characterises only a single terminal taxon – an autapomorphy (Text-fig. 34).

In the example shown in Text-fig. 34, both the synapomorphy and the autapomorphy have a consistency index of 1 – each has a theoretical

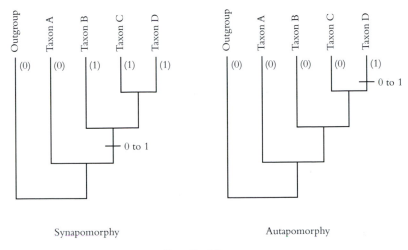

Synapomorphy   Autapomorphy

Text-fig. 34

minimum number of steps of 1, and each changes only once on the tree. So the ci for both of these characters would be 1. But they are not equally useful. In this case, the synapomorphy defines a clade including Taxa B, C and D, while the autapomorphy defines only Taxon D. In other words, because they are shared derived features, synapomorphies are informative, whereas autapomorphies, being unique, are not.

Essentially what distinguishes these two characters is the degree to which they are 'retained' by succeeding taxa up the cladogram. The synapomorphy is common to Taxa B, C and D following the transformation of that character at the base of that clade. It is retained by all the succeeding taxa. By contrast, the autapomorphy is not shared by any taxa other than D, so is not retained by any succeeding taxa. The *retention index*, or *ri*, is a measurement of this retention of character transformations:

$$\text{Retention index} = \frac{\text{Maximum number of steps} - \text{Observed number of steps}}{\text{Maximum number of steps} - \text{Theoretical minimum number of steps}}.$$

The maximum number of steps is the number of steps a character would show if a different state for any given character were acquired independently by each taxon that shows that character state. It is the number of steps that the character would have on the least parsimonious conceivable tree – though with the proviso that by convention one chooses the smaller number of taxa showing a different state. Text-figure 35 plots the same characters as shown in Text-fig. 34, but this time notice the different

59

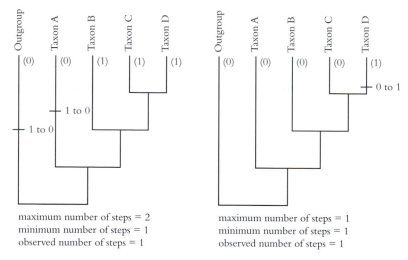

maximum number of steps = 2
minimum number of steps = 1
observed number of steps = 1

maximum number of steps = 1
minimum number of steps = 1
observed number of steps = 1

Text-fig. 35

values of maximum number of steps for each character. The one on the left, which was the synapomorphy for the clade including Taxa B, C and D, could transform twice, to yield the 0 for the outgroup, and for A. By contrast, the character on the right, which was an autapomorphy of Taxon D, can be transformed only once.

So, for the synapomorphic character, on the left, the maximum number of steps is two. The minimum number of steps is one, since there is only one possible character transformation, from 0 to 1, and the observed number of steps is one. Therefore the retention index can be calculated as:

$$\text{Retention index} = \frac{(2-1)}{(2-1)} = \frac{1}{1} = 1.$$

For the autapomorphic character, shown on the right, the maximum number of steps is one, and the minimum number of steps is also one. The observed number of steps is one, and so the retention index can be calculated to be 0/0:

$$\text{Retention index} = \frac{(1-1)}{(1-1)} = \frac{0}{0}.$$

For our purposes we can treat the value 0/0 as being zero, though strictly speaking anything divided by zero is an undefined number. But since this character has 'no retention' up the cladogram, by convention a value of zero is used. The value for retention index always ranges from 0 and 1. Retention index values of 0 occur where there is 'no retention' of a transformation along a tree, for example when the transformation is confined to a single taxon branch, or to transformations occurring on single taxon branches on remote parts of the tree. By contrast, a retention index of 1 will occur where a transformation on a branch leading to a clade is retained by all the succeeding branches within that clade. The retention index is therefore a useful indicator of whether or not a character is likely to be phylogenetically useful, as it increases in value the more common a transformation is to all the taxa within a clade. In practice, characters with high retention indices are also worth looking at closely to see if they can be used to help define clades.

Screens 22/33 to 33/33

Look at the distribution of the character transformations in the data matrix on Screen 22/33. Character 1 exhibits the same state (0) in the outgroup taxon X and all the ingroup taxa except E. This character is therefore an autapomorphy of Taxon E. The maximum number of character transformations conceivable for this character (g) is one, as is the minimum number (s). Naturally, the observed number of transformations (s) for this character is one as well. The retention index can be calculated to be $(g - s)/(g - m) = (1 - 1)/(1 - 1) = 0/0$, which we take to be zero.

Character 2 apparently went from state 0 to state 1 independently in Taxa A and B. Therefore, it is not a synapomorphy of these two taxa, but rather an example of homoplasy. The maximum number of steps is two, while the minimum number would be one if A and B formed a clade with the transformation of character 2 from 0 to 1 as its synapomorphy. The observed number of steps is two, so that the retention index can be calculated thus: $(2 - 2)/(2 - 1) = 0/1 = 0$. Note that both this homoplasious character and the previous autpomorphic one score retention indices of zero.

Character 3 has three states, 0, 1 and 2. Taxon A has state 1 while Taxa D and E have state 2. All the other taxa have state 0. State 1 for A is autapomorphic since it is unique to that taxon and so does not appear to be an intermediate leading to character state 2, i.e., character states 0, 1 and 2 do not make up an ordered sequence of states. The maximum possible number of character state transformations for this character is three. There is one transformation from 0 to 1 for Taxon A, and two independent transformations from 0 to 2 for D and E. The minimum number of steps is two, and the observed number of character transformations is two. The retention index for this character is therefore $(3 - 2)/(3 - 2) = 1/1 = 1$.

The fourth character is a bit tricky (Screen 26/33). There are two ways of interpreting the distribution of character states here. One is to say that character state 0 is the derived state and that it evolved independently in the outgroup and in Taxon B, shown on the screen diagram by the dark blue bars. The alternative is that state 0 is primitive for the entire group, and that the transformation from 0 to 1 occurred independently in Taxon A and in the clade including C, D and E, as shown on the diagram by the pale blue bars. Taking the first hypothesis, the maximum number of steps (g) would be two, i.e., once from 1 to 0 for the outgroup X and the same transformation once again for Taxon B. By contrast, the second hypothesis yields a maximum number of steps of four, i.e., a transformation from 0 to 1 for A and then the same transformation three times for the branches leading to C, D and E. By convention, the hypothesis which produces the smallest value for the maximum number of steps is used. Since there are two character states, the minimum number of steps is one, and the observed number of steps is two. Therefore the retention index is $(2 - 2)/(2 - 1) = 0/1 = 0$.

The fifth character is clearly a synapomorphy of the clade including D and E. The maximum number of steps would therefore be two (independent character transformations for D and E). Again, this is a binary character, so the minimum number of steps is one and the observed number of steps is one. The retention index is therefore $(2 - 1)/(2 - 1) = 1/1 = 1$.

Character 6 is a synapomorphy of the clade including C, D and E. Either the group C, D and E underwent independent transformations from 0 to 1, or they all retain the character state 1 from a single transformation at the base of that clade. Hence, the maximum number of steps for the character is three and the minimum number is one. Since the observed number of steps is one, the retention index is $(3 - 1)/(3 - 1) = 2/2 = 1$.

Character 7 is an autapomorphy of Taxon C. The maximum number of steps is one, the minimum number is one, and so the retention index for this character is zero, as was the case for character 1, which was also an autapomorphy.

Character 8 shows a parallel transformation from 0 to 1 in both A and the clade including D and E, i.e., two steps. The maximum number of steps for this character would therefore be three. The minimum number of steps is one, and the observed number of steps is two. The retention index for this character is $(3 - 2)/(3 - 1) = 1/2 = 0.5$.

Homoplasy is again evident with character 9, where there appear to be parallel transformations from state 0 to state 1 for Taxa A and B, while the rest of the ingroup taxa retain the same state as the outgroup. The maximum number of steps is two, the minimum is one, and the observed number is two. Thus the retention index for this character is $(2 - 2)/(2 - 1) = 0/1 = 0$.

By contrast, character 10 is the synapomorphy of the clade including the Taxa B, C, D and E. As with character 4, there are two ways of explaining the pattern of characters seen here. Taxa B, C, D and E could all have independently acquired the transformation 0 to 1 for this character, but a more parsimonious explanation is that the outgroup X and Taxon A independently gained the transformation 1 to 0. Thus, there is a maximum number of steps of four for the first explanation and only two for the second. In calculating the retention index remember that the smallest value is used. Since the minimum number of steps is one, as this is a binary character, and the observed number of steps is 1, the retention index comes out as $(2 - 1)/(2 - 1) = 1/1 = 1$.

An ensemble retention index for the tree can be calculated (Screen 29/33). This can be taken as an indication of how much homology compared with homoplasy there is in the cladogram, i.e., how much phylogenetic signal these characters are reflecting. The ensemble retention index for the tree is:

$$\text{Ensemble retention index } (RI) = \frac{(G - S)}{(G - M)},$$

where G is the sum of all the maximum numbers of steps for each character, M is the sum of all the minimum numbers of steps, and S is the sum of all the observed numbers of steps. High retention index values, close to 1, imply minimal homoplasy and a cladogram which retains a strong phylogenetic signal. The observed branching of the tree could be viewed with confidence as a real reflection of the phylogeny of the group and of the evolution of the characters. Low retention index values, close to 0, mean the opposite: the cladogram is dominated by homoplasy, with many of the arrangements of taxa brought about by chance, convergence or parallelism. Such a tree would have to be used with caution in determining the phylogeny of a group (Screen 31/33).

An additional use of the retention index is in re-weighting characters, a process whereby homoplasious characters are 'handicapped' so as to minimise their effect on subsequent analyses of the tree. Note that this is a very different proposition to the *a priori* weighting advocated by some biologists. While *a priori* weighting uses preconceptions to re-weight characters before the analysis has started, this *a posteriori* re-weighting is strictly empirical. The first analysis does not re-weight the characters, and re-weighting is done afterwards according to the preliminary results. The *rescaled consistency index* is used to re-weight the characters. This is simply the product of the consistency index and the retention index.

4.2 – Tests of robustness

Screens 1/8 to 8/8

Although knowing how informative certain characters are is useful, it is only part of the process of interpreting a cladogram: the topology of the cladogram is also important. It is the topology that describes the relationships of the taxa, i.e., the arrangement of nodes. Not all nodes are equal: some may be defined by only a single character transformation, while others may be defined by many. Another property of nodes is *robustness*. This can be thought of as how resistant a node is to changes in the data matrix. A robust node will remain in place even if there is significant rearrangement or resampling of the characters in the data matrix, while less robust nodes will not. Many biologists consider robustness to be a good indicator of the usefulness of a node in describing the phylogeny of a group of organisms. However, since tests for robustness are statistical rather than biological, there is no fundamental reason why a weak node shouldn't better represent historical reality than a robust one. What tests of robustness do indicate is the likelihood of such nodes appearing in the most parsimonious cladogram simply through chance. There are two widely used robustness tests, the *bootstrap test* and the *Bremer support test* (sometimes known as the branch decay test).

Bootstrapping is performed by randomly resampling the data matrix through duplicating some characters while removing others, though preserving the same number of characters as in the original data matrix. A parsimony analysis is then performed, and the new most parsimonious tree is compared with the one from the original data matrix to see which nodes they have in common. Usually this is done on a computer, where a large number of such resampled data matrices (or replicates) can be quickly produced and processed, though the bigger the data matrix the longer it takes to do. Nodes that are seen in the most parsimonious tree from the original data matrix and in a large proportion of the trees from the resampled data matrices are deemed to be robust. Such nodes are defined by many characters, and so are tolerant of changes to the data matrix. The results are usually presented as a consensus tree (Screen 4/8), commonly with only those branches present in more than half of the bootstrap trees being shown.

Bremer support testing is a measurement of how persistent nodes in the most parsimonious trees are in successively less parsimonious trees. As shown on Screen 5/8, the first step is to run the computer analysis to identify the length of the most parsimonious tree or trees. Now the computer is set to with a new 'constraint' – not to find the most parsi-monious tree, but to find all the trees exactly one step longer. A strict consensus tree is made, and the nodes in common between the original tree and this consensus tree are recorded. This is repeated, this time with the trees of two steps longer, then three steps, and so on until the strict consensus tree is a single polytomy, i.e., all the nodes have collapsed (Screen 7/8). The Bremer support index is the number of extra steps it takes to collapse a node, and the more steps, the greater its robustness.

It is common for biologists to perform both bootstrap and Bremer support tests on their data sets. These tests are desirable because they offer a handle on the usefulness or otherwise of the cladogram deduced from a given data set. The higher the bootstrap and Bremer support indices, the more robust the data set is deemed to be, and so the more confidence can be placed in a phylogeny deduced from it.

5–Practical

5.1 – Phylogenetic analysis of eight species of sea-urchins

Screens 1/2 and 2/2

In the preceding part of this booklet I have tried to supplement the notes on the cladistic process presented on the CD-ROM with additional details and examples. Where necessary, I have written step-by-step explanations to answering the questions and puzzles set before you. For the final section, however, such an approach is not really appropriate if the exercises laid out for you here are to be truly useful. Instead I will try to summarise what it is you are being asked to do at each stage – but the rest will be all your own efforts! Finally, do remember to use the glossary feature on the CD-ROM to look up technical terms quickly (the question mark button on the **functions** menu at bottom right).

The aim of this practical is to determine the phylogeny of eight species of sea–urchins, first using morphological characters and then using molecular data. Sea–urchins, or echinoids, are an important and diverse group of Echinodermata, a large phylum of marine animals that also includes the starfish, brittle-stars, sea cucumbers and crinoids. They are characterised by an endoskeleton, or test, made up of calcitic plates. Attached to the test are a variety of spines, arms and suckers. Sea–urchins are commonly divided into two main groups – the regular sea–urchins and the irregular sea–urchins. The regular sea–urchins have more or less globular tests exhibiting clear radial symmetry. Generally they have well-developed spines, giving them a very spiny appearance. The mouth is on the ventral surface and is equipped with a ring of five sharp teeth for scraping algae and small animals from the sea floor. The opening of the anus, or periproct, is on the dorsal surface of the test. Irregular sea-urchins are not radially symmetrical and instead have a definite front and back, and the spines are often smaller and finer, giving the animal a furry rather than a spiny appearance. The mouth is on the ventral surface, but instead of having scraping teeth the sea-urchin swallows detrital particles and extracts the nutrients internally. The periproct is not on the dorsal surface but towards the back of the animal, at or close to the ventral surface. On the whole regular sea-urchins live on the sediment, while irregular sea-urchins burrow through it.

An interesting aspect of this practical is that you will get to compare the morphological and molecular cladograms produced for the same set of organisms. Although the morphological attributes of an organism (the phenotype) and its molecular, genetic code (the genotype) are intimately related, they are totally different things as far as characters are concerned. In a way, they each offer unique sources of data for reconstructing phylogeny. Therefore if they both present essentially similar stories as far as the phylogeny of a group of organisms is concerned, one can be much more confident about having discovered 'the truth'. If, on the other hand, they are significantly at variance then one must approach both phylogenies with much more caution. For this reason, combining morphological and molecular phylogenetic analyses can be one of the best ways to study the evolutionary history of a group.

5.2 – Cladistic analysis of morphological characters

Screen 1/5

On the first screen of this section notice the list of taxa down the left-hand side, and the characters along the top. The characters are numbered 01, 02, and so on to 35. There are four options along the bottom of the screen, three of which will be available at any given time: **View Matrix**, **Browse Database**, **View Completed Matrix**, and **Instructions**. Clicking on these will take you to a different screen, and by default you should start this section at the **Instructions** screen. Initially the **View Matrix** screen will show you an empty matrix, but as you work through this exercise you will fill it in. The **View Completed Matrix** is what the matrix should look like after all the data have been entered. The **Browse Database** screen is where you can find out about the taxa and characters, and so compile the character matrix. Go to the **Database** screen by clicking on

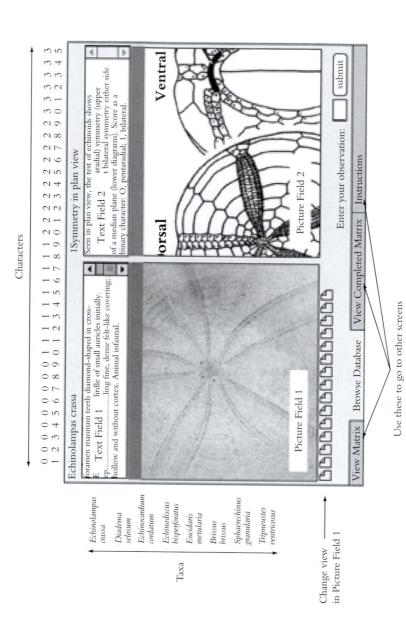

Text-fig. 36

the **Browse Database** button. Next, click on any of the taxa on the left to select one. Clicking on the character numbers at the top selects the descriptions and diagrams relevant to that particular character. Note that you can fill in the matrix bit by bit. If you quit the application, or turn off the computer, the data matrix will store your entries on your computer's hard disk. So you don't need to do the whole thing in one sitting!

To begin with, choose *Echinolampas crassa*, the first echinoid on the list, and then choose the first character, 01. You should be presented with a screen like that shown in Text-fig. 36.

Notice that this screen contains four fields, two text fields at the top, and two picture fields beneath them. The text field on the left (labelled here as Text Field 1) gives a basic account of this particular sea-urchin. This is very similar to the sorts of description found in taxonomic monographs, which describe species in a very precise, terse and often rather formulaic manner. Beneath this text field is a picture of the test, or shell, of the sea-urchin (Picture Field 1). By clicking and holding the mouse over this picture you can drag the image around, allowing you to view regions outside the initial field of view. Immediately underneath this picture is a row of document icons. If you click these the picture in the field changes. The first icon brings up the top view, the next the bottom view, and so on. At top right is another text field (Text Field 2), which contains a specific and detailed description of each character. The content of this field changes if you click on the character numbers at the top. So clicking on 01 brings up a description of the symmetry of this particular sea-urchin, 02 a description of the shape of the test, and so on for all thirty-five characters. To see these features on the sea-urchin in question, you will need to scan through the different views available in Picture Field 1. To help you understand some of the descriptions, there is a diagram in Picture Field 2. As with Picture Field 1, you can click and drag the diagram about. At the top is a regular (radially symmetrical) sea-urchin, and below is an irregular (bilaterally symmetrical) one. Some key features are labelled. Underneath Picture Field 2 is a box for entering the character state you deem to be appropriate for a certain character on a certain species. When you enter a character state here and click on the submit button, it is sent to the screen with your data matrix on it. To show you how this is done, we will run through the first species, *Echinolampas crassa*, together. Note that when you submit each character state you will be taken to the data matrix. To return, click on the **Browse Database** button.

1. Symmetry in plan view: Echinoids exhibit either pentaradial symmetry, as in a starfish, or bilateral symmetry, as in ourselves. *Echinolampas crassa* is not radially symmetrical — it has a definite front and back. The ventral view shown in Picture Field 1 shows this clearly (click on the second document icon to bring this picture up). Note that the periproct is on this, the ventral, surface, similar to the arrangement in the irregular sea-urchin shown in Picture Field 2. By contrast, regular sea-urchins have the periproct on the dorsal surface. So, for this character, the correct character state is 1 — type a '1' into the character state box at the lower right and click on **submit**. Note that this character is a binary character as there are only two options.

2. Test shape: Despite their morphological variety, sea-urchins come in four main shapes. Most of the regular sea-urchins and some irregular ones are more or less globular, or spherical, in shape. Irregular sea-urchins are sometimes strongly flattened, like the sand dollars we looked at earlier (Section 3.1). Others are ovate, i.e., shaped like an egg. Yet others have a very distinctive heart shape, with two lobes at the front, and with the sides tapering backwards to meet at the rear. *Echinolampas crassa* has an ovate shape, i.e., character state 2. Note that this is an unordered multistate character: it has more than two options, but there is no evidence, *a priori*, to make them an evolutionary sequence.

3. Periproct position – 1: The periproct is either enclosed by the apical disk (the discrete ring of plates at the dorsal apex of the test) or it is not. In *Echinolampas crassa* it is not, so code this character as 1. Such 'either/or' options are typical binary characters.

4. Periproct position – 2: The periproct can be found on the dorsal surface, the posterior of the test, or underneath the animal on the ventral surface, depending on the species. If you look at the close-up view of the apical disk of *Echinolampas crassa* you can see that it is located at the posterior edge of the ventral surface. Therefore, character state 2 should be selected.

5. Apical disk – plate arrangement: The apical disk is the distinctive ring of plates at the apex of the test. The periproct may or may not be enclosed by the apical disk. There are three options here. First, the genital and ocular plates can alternate with one another to form the ring. This is the monocyclic condition. Alternatively, the genital

and ocular plates can form more or less distinct rings, the dicyclic condition. Finally, the plates can be compacted or fused into a single large plate. *Echinolampas crassa* exhibits this final condition, with the genital plates fused into a single large plate, and the five small ocular plates surrounding it (look at the apical plate diagram). So this character is coded as 2.

6. Apical disk – number of gonopores: This is another simple binary character. The two options are either five gonopores or four gonopores. *Echinolampas crassa* has four gonopores and so the correct character state code here is 1.

7. Apical disk – number of genital plates: Depending on the species, there can be five genital plates, four genital plates, or just one large plate. The last condition is the case with *Echinolampas crassa*, and so the correct character coding is 2.

8. Lunules: These are the slots seen in the tests of some sea-urchins, such as the sand dollars we looked at in Section 3.1. *Echinolampas crassa* lacks them, and so scores 0 here.

9. Ambulacral structure: The ambulacral plates may be simple or compound. Simple ambulacral plates are small and bear a single spine base, or tubercle. They also have a single pore pair. Compound plates consist of two or more plates united into a single structure and bearing one large spine base. There are two versions of compound plating. Diadematoid plating consists of a central large plate with small plates above and below, while echinid plating has the largest plate at the bottom and a number of small, wedge-shaped ones above it. Look at the ambulacral plating diagrams to see these. In particular, compare the diadematoid plating of *Diadema setosum* with the echinid plating of *Tripneustes ventricosus*. Compound plates can have several pore pairs. *Echinolampas crassa* has simple plating, and so scores 0.

10. Petals: Where the ambulacra become modified as respiratory organs, there are modified plates and pores which form distinctive petals running from the apical disk. This is the case with *Echinolampas crassa*, as can be observed in the apical view of the test. Therefore this character should be coded as 1.

11. Frontal ambulacrum: In some species the frontal ambulacrum has modified plates and so looks different when compared with the other four, e.g., look at the apical view of *Brissus brissus*. This is not the case with *Echinolampas crassa* and so this character can be coded as 0.

12. Ambulacral pores – number per ambulacral plate: There is either a single pair of openings in each ambulacral plate or a number of minute pairs. Be careful not to confuse the spine bases for the openings! On the ambulacral plating diagram the openings are shaded black. *Echinolampas crassa* clearly has a single pair, and scores 0 for this character.

13. Ambulacra – oral tube-feet: *Echinolampas crassa* is stated in the species description to have suckered not penicillate ('adoral') tube-feet around the mouth, so this character scores 0.

14. Ambulacral pores – adoral pores: These are either paired (Picture Field 2, fig. 1) or single (Picture Field 2, fig. 2). The close-up view (the 'pores (oral)' close-up view) shows that on *Echinolampas crassa* they are single, so choose character state 1 for this character.

15. Tuberculation – 1: There may be a single large tubercle on each ambulacral plate (Picture Field 2, fig. 2) or a number of more or less similar-sized, large tubercles (Picture Field 2, fig. 3). In *Echinolampas crassa* the latter is the case, so choose character state 1 for this character.

16. Tuberculation – 2: Depending on the species the tubercle may be perforate or imperforate. The tubercles of *Echinolampas crassa* are quite obviously perforate, with a hole running down the centre. Choose character state 0 for this character.

17. Tuberculation – 3: Another variation in the tubercles is the shape of the platform surrounding the central boss. It may be either smooth or crenulate. The tubercles of *Echinolampas crassa* are crenulate, with the 'cogs' running outwards from the central boss. Therefore, choose character state 1 for this character.

18. Fascioles: These are the narrow bands which cover certain parts of the test in some species. In the example shown in Picture Field 2 they are the zigzagging grooves. *Echinolampas crassa* does not appear to have them, so it scores 0 for this character.

19. Peristome – 1: This character is based on the size of the structure. The peristome is said to be either large or small, depending on its size in relation to the overall size of the test. In *Echinolampas crassa* the peristome is small, and so scores 1.

20. Peristome – 2: There may or may not be buccal slits (or notches) around the peristome. In Picture Field 2 compare the sea-urchin at the top, which has these slits, with the sea-urchin at the bottom, which does not. *Echinolampas crassa* does not have them either, and

so scores 0. Be careful not to confuse the five slightly recessed ambulacra for buccal slits, which are notches in the interambulacra (look carefully at the peristome view in Picture Field 1).

21. Peristome – 3: The peristome may be either central on the lower surface (as can be seen in the upper figure) or markedly displaced towards the anterior (as in the middle figure). It is almost central in *Echinolampas*, which you should thus score as 0 for this character.

22. Spines – 1: These are divided into two types – those lacking a cortex (Picture Field 2, fig. 2) and those possessing a cortex (Picture Field 2, fig. 1). The spines lacking a cortex are relatively open, the edges folding in and out, and in life are enclosed by skin. By contrast, the spines with a cortex are enclosed in a calcitic sheath, the cortex. If you look at the spines of *Echinolampas crassa* (Picture Field 1, spine structure view), you can see they lack a cortex. Therefore, code this character 0.

23. Spines – 2: The spines may either have a large central cavity, or lumen (Picture Field 2, fig. 1), or be more or less solid (Picture Field 2, fig. 2). The spines of *Echinolampas crassa* (Picture Field 1, spine structure view) have the central lumen, so code this character 0.

24. Spines – 3: The distribution of spines is very variable, but three basic arrangements can be recognised: short and spiky, short and felt-like, and curved and hair-like. The spines on *Echinolampas crassa* are stated in the species description to be felt-like, so enter a 1 for this character.

25. Spines – 4: In some species the spines form a grill over the mouth, in others they do not. The spines on *Echinolampas crassa* do form such a grill, as shown in Picture Field 1, spine cover (oral) view. So for this character a 1 can be entered.

26. Sphaeridia: These are either present or absent. According to the species description they are present on *Echinolampas crassa*, and so a 1 can be entered for this character.

27. Internal structures – perignathic girdle: Auricles may or may not be present. In the species description it is stated that in *Echinolampas crassa* they are present in juveniles, but degenerate as the animal matures. But since they are present, albeit briefly, score this as 1.

28. Internal structures – perignathic girdle: Apophyses – high plates associated with the jaw apparatus – may or may not be present. In *Echinolampas crassa* they are absent, as can be seen in the perignathic girdle view in Picture Field 1. So score this as 0.

29. Internal structures – buttressing: In certain species of sea-urchin the inside of the shell is reinforced with struts and columns. These are not present in *Echinolampas crassa*, and so for this character enter 0.

30. Lantern – 1: This is either absent, present only in juveniles, or present throughout the life of the animal. In the species description for *Echinolampas crassa* it is stated that the lantern is present at first but lost in adults, so for this character the correct score is 1.

31. Lantern – 2: The teeth are either U-shaped, keeled, or diamond-shaped in cross-section. In the species description for *Echinolampas crassa* it is stated that the teeth are diamond-shaped in cross-section, so the correct score is 2. Note that this is a character where a '?' should be used where teeth are not present.

32. Lantern – 3: The lantern is either tall, with upright teeth, or depressed, with nearly horizontal teeth, depending on the species. From the species description for *Echinolampas crassa* it is evident that the lantern is of the tall, vertically mounted arrangement, so the correct score is 0. Note that once again a '?' might need to be used for species lacking a lantern.

33. Lantern – 4: The region where the teeth are not touching is called the foramen magnum. In some species of sea-urchin this foramen magnum is relatively shallow, 0 to 10% of the height of the teeth (Picture Field 2, fig. 2); in others it is quite deep, 10 to 50% (Picture Field 2, fig. 3). If you look at the diagram of the lantern of *Echinolampas crassa* (Picture Field 1, lantern view) you can see that the five teeth meet along the lower surfaces but diverge from about the midpoint upwards, and so this character should be scored 1. Once again a '?' might need to be used for species lacking teeth.

34. Lantern – 5: Epiphyses may be separate or they may be in contact at some point above the foramen magnum. If you look at the diagram of the lantern of *Echinolampas crassa* (Picture Field 1, lantern view) you can see that the epiphyses do not meet, and so this character should be scored as 0. Once again a '?' might need to be used for species lacking a lantern.

35. Plastron: This is the large modified posterior interambulacrum present on the ventral surface in certain sea-urchins. *Echinolampas crassa* (Picture Field 1, ventral view) lacks a plastron, all the ventral interambulacra being rather similar, and so this character should be scored as 0.

Now repeat this process with all the other sea-urchins to complete the data matrix.

Screens 2/5 to 5/5

When the character matrix is analysed, using *Eucidaris metularis* as the outgroup, two most parsimonious trees are found (Screen 2/5). The strict consensus tree contains a four-way polytomy, with one branch leading to the outgroup, another to *Diadema setosum*, a third to a clade including *Sphaerechinus granularis* and *Tripneustes ventricosus*, and a fourth to a clade including *Echinolampas crassa*, *Echinocardium cordata*, *Brissus brissus* and *Echinodiscus bisperforatus* (Screen 3/5). A bootstrap test of the data provides an interesting comparison with these trees. There is very strong support for the *Sphaerechinus granularis* and *Tripneustes ventricosus* clade and the *Echinolampas crassa*, *Echinocardium cordata*, *Brissus brissus* and *Echinodiscus bisperforatus* clade (99% for the former, and 100% for the latter). However, the support for *Diadema setosum* as the sister group of the *Echinolampas crassa*, *Echinocardium cordata*, *Brissus brissus* and *Echinodiscus bisperforatus* clade is very low – a marginal 51%.

5.3 – Cladistic analysis of molecular characters

Screens 1/10 to 10/10

At first glance, the molecular data set which can be brought up on Screen 1/10 looks incredibly complicated, but the thing to remember is that most of the bases are the same, and that what you're looking for are the few small differences that distinguish the sequences from one another. The trick is to spot the long similar sequences that can be used to link sequences together, and conversely, the bits of sequences that are different. Don't get distracted by single bases that are different, particularly when one pyrimidine has been replaced by the other, or one purine with the other. To help you, we'll go through the first three taxa together. Remember that after getting two sequences to 'match', adding the additional species is likely to demand further adjustments; for example, there might be duplications of sequences in the other species. With every additional species you need to run through the previous species again to check for such adjustments.

We'll begin with *Eucidaris metularis* and *Diadema setosum*. Click the buttons to enter their molecular sequences. Looking at the first fifty

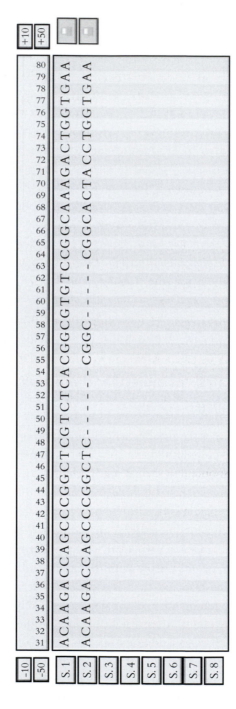

Text-fig. 37

bases for these two species, it is immediately apparent that the first forty-seven bases are the same. The next long series in common is CGGC, bases 55 to 58 in *Eucidaris metularis* and 49 to 52 in *Diadema setosum*. So insert six dashes after position 48 in the *Diadema setosum* sequence to line these up from position 49. Now there are five bases that match, CCGGA, bases 64 to 68 in *Eucidaris metularis*, 59 to 63 in *Diadema setosum*. Insert five dashes in the *Diadema setosum* sequence from position 59 to line these up. Once you've done this, you'll notice that all the remaining bases match more or less perfectly (Text-fig. 37).

Now add a third taxon, *Sphaerechinus granularis*. Notice that the seventh to forty-seventh bases of the first two taxa match almost perfectly the eleventh to forty-eighth bases of *Sphaerechinus granularis*. You need to insert a dash at the seventh position on the sequences of *Eucidaris metularis* and *Diadema setosum* to make them line up with *Sphaerechinus granularis*. Note that if you add two dashes, so as to make locations 9 and 10 match, all the others stop matching. The next two changes are similar to the ones *Diadema setosum* needed. First, you need to insert six dashes from position 50 to match up the sequence CGGC with that at positions 56 to 59 of *Eucidaris metularis* and *Diadema setosum*. Then five dashes from position 60 to align the CGGTA sequence with the same sequence for the first two species. If you look at the bases from position 74 for *Sphaerechinus granularis* you can see that there is a long series of them which match the bases from position 75 for the first two species (TGGT...). Add a dash at position 74 to bring them into alignment. Finally, if you insert a dash at position 269 in the *Sphaerechinus granularis* sequence, you will bring all the remaining bases into correspondence with those of the other two species.

The rest of the exercise is a continuation of what we have done here. Don't forget that you may need to adjust these first three species – they are not necessarily finished with. Sometimes you need to adjust the new species to match the others, and sometimes the other species to match the new ones. Once you have done this you need to decide which characters are informative (see Section 2.3). Remember, you are trying to find characters which might unite two or more taxa into clades, while excluding others. For example, character states unique to one species (autapomorphic, phylogenetically neutral sites) are no good, e.g., base positions 34 and 50. Neither are character states shared by all eight species (invariant sites), such as base positions 1 to 5. Eventually, you should end up with twenty-two informative characters.

Once again two most parsimonious trees are obtained (Screen 7/10). This is emphatically *not* a property of molecular and morphological data sets of the same groups of species — there is no reason why they should have the same number of most parsimonious trees. A strict consensus tree can be made from these (Screen 8/10), and the data set can be subjected to a bootstrap test (Screen 9/10).

5.4 – Comparison of results and conclusions

Screens 1/9 to 9/9

When the most parsimonious trees of the morphological data set (Section 5.2) and the molecular data set (Section 5.3) are compared, some differences and some similarities can be seen. The easiest way to highlight these is to make a strict consensus tree, which will retain only features they have in common (Screen 2/9). Four species comprise a clade common to all four cladograms — *Echinolampas crassa*, *Echinocardium cordata*, *Brissus brissus* and *Echinodiscus bisperforatus*. What these four species have in common is that they are all infaunal, meaning they burrow through the sediment (see the species descriptions in Section 5.2). It could be argued that morphological features might have evolved convergently as a result of these four species belonging to groups which independently became infaunal, in which case these similarities would be a convergent homoplasy. However, since the same clade is evident from the molecular data, it does seem that the group is real. Molecular similarities are unlikely to have been subjected to the same convergent selection pressures from a burrowing lifestyle as, say, the shape of the test or the morphology of the spines (Screen 3/9). A further stage in the analysis is to combine both the molecular and the morphological data into one big data set, and compare the resulting most parsimonious tree or trees with those of the separate data sets (Screen 5/9). Look to see whether the nodes in the combined data set tree match those in the trees from the separate data sets. In cases of conflict, the bootstrap values can be useful in deciding which alternative has greater probability.

Appendix: Cladistics software

Cladistics leans heavily on the use of computers. There are a number of websites devoted to the discussion and distribution of applications useful to biologists doing cladistic analyses. Some of these applications are commercial products, but many can be downloaded and used for free ('freeware'). Applications for both the Macintosh and Windows operating systems are widely available. Pages with links to many of the best pieces of cladistic software can be found at the sites listed below.

General cladistics software review and link pages

- The Taxonomy and systematics archive at the University of Glasgow:
    http://taxonomy.zoology.gla.ac.uk/software.html
- Willi Hennig Society education and software page:
    http://www.cladistics.org/education.html
- Indiana University molecular biology software archive:
    http://ftp.bio.indiana.edu/IUBio-Software+Data/molbio/evolve/

Parsimony analysis applications

The three most widely used applications for analysing data sets and searching for trees are PAUP, HENNIG86 and PHYLIP. The MacOS version of PAUP is by far the easiest to use.

- *PAUP (Commercial)*
  Published by Sinauer Associates (http://www.sinauer.com/)
  Currently version 3.x MacOS only, version 4 betas MacOS and Windows.
    http://paup.csit.fsu.edu/index.html

- HENNIG86 (Commercial)
  Distributed by Arnold Kluge, University of Michigan (akluge@umich.edu) and
  Diana Lipscomb, George Washington University (biodl@gwuvm.gwu .edu).
  MS-DOS only.
- *PHYLIP (Freeware)*
  http://evolution.genetics.washington.edu/phylip.html

Analysis of unrooted networks

- *SplitsTree 2.0 (Freeware)*
  Written by Daniel Huson (huson@member.ams.org)
  MacOS, Windows and UNIX.
  http://www.mathematik.uni-bielefeld.de/~huson/phylogenetics/
  splitstree.html

Tree analysis

- *MacClade (Commercial)*
  Published by Sinauer Associates (http://www.sinauer.com/)
  MacClade is the most widely used program for putting together character
  data matrices. It is often used with PAUP with which it integrates very well.
  It will also import trees from PAUP and can be used to make detailed analyses
  of tree shape, character state changes, etc. MacOS only.
  http://phylogeny.arizona.edu/macclade/macclade.html
- *TreeView (Freeware)*
  Written by Rod Page (r.page@bio.gla.ac.uk)
  A useful application for viewing and printing trees. MacOS and Windows.
  http://taxonomy.zoology.gla.ac.uk/rod/treeview.html

  Note: The publisher has used its best endeavours to ensure that the URLs for external
  websites referred to in this booklet are correct and active at the time of going to press.
  However, the publisher has no responsibility for the websites and can make no guar-
  antee that a site will remain live or that the content is or will remain appropriate.